War Without Winners

Afghanistan's Uncertain Transition After the Cold War

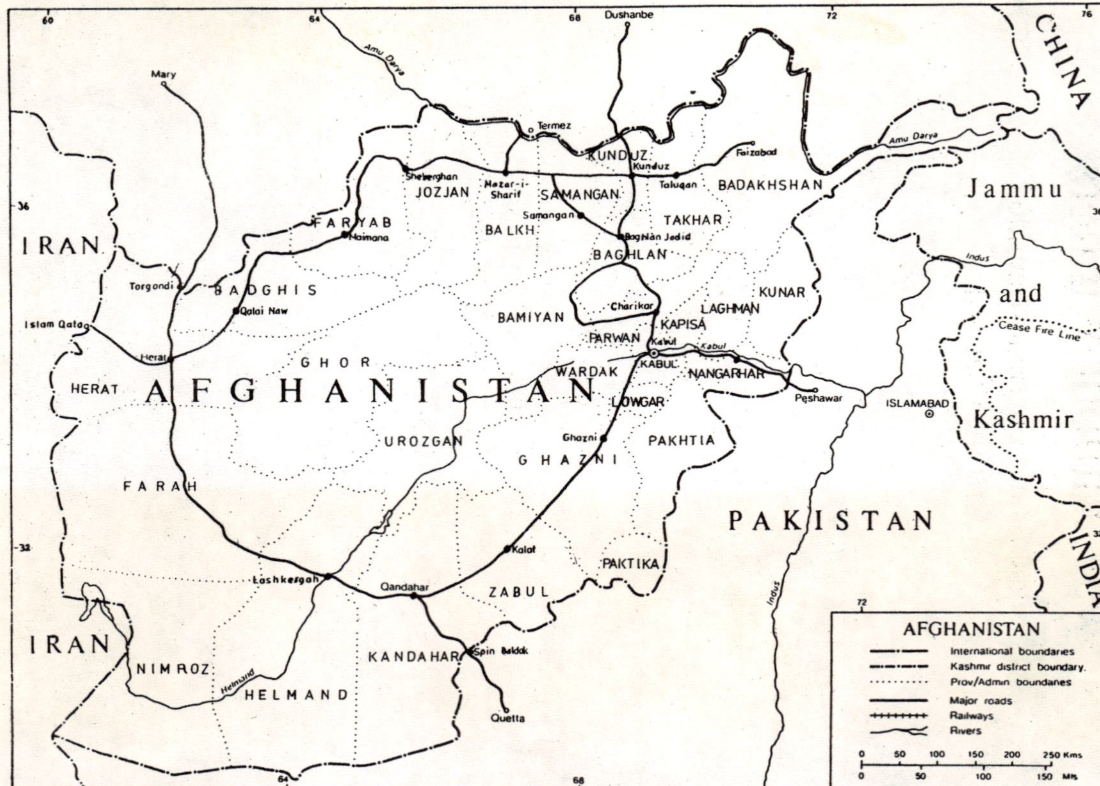
AFGHANISTAN
CHINA
Jammu
and
Kashmir
Cease Fire Line
INDIA
PAKISTAN
IRAN
IRAN
Dushanbe
Termez
Mary
Amu Darya
Amu Darya
Indus
Indus
Kabul
KUNDUZ
Kunduz
Faizabad
Taluqan
BADAKHSHAN
TAKHAR
Sheberghan
JOZJAN
Mazar-i-Sharif
SAMANGAN
Samangan
BALKH
Baghlan Jadid
BAGHLAN
FARYAB
Maimana
Torgondi
BADGHIS
Qalai Naw
Islam Qala
Herat
HERAT
KUNAR
Charikar
LAGHMAN
BAMIYAN
KAPISA
PARWAN
Kabul
KABUL
NANGARHAR
Peshawar
ISLAMABAD
GHOR
WARDAK
LOWGAR
UROZGAN
Ghazni
GHAZNI
PAKHTIA
FARAH
Kalat
PAKTIKA
Lashkergah
Qandahar
ZABUL
KANDAHAR
Spin Baldak
Quetta
NIMROZ
Helmand
HELMAND
60
64
68
72
76
36
32
AFGHANISTAN
International boundaries
Kashmir district boundary.
Prov/Admin boundaries
Major roads
Railways
Rivers
0 50 100 150 200 250 Kms
0 50 100 150 Mls

War Without Winners

Afghanistan's Uncertain Transition After the Cold War

Rasul Bakhsh Rais

Karachi
Oxford University Press
Oxford New York Delhi
1994

Oxford University Press, Walton Street, Oxford OX2 6DP
OXFORD NEW YORK TORONTO
DELHI BOMBAY CALCUTTA MADRAS KARACHI
KUALA LUMPUR SINGAPORE HONG KONG TOKYO
NAIROBI DAR ES SALAAM CAPE TOWN
MELBOURNE AUCKLAND MADRID
and associated companies in
BERLIN IBADAN

OXFORD is a trade mark of Oxford University Press

First edition 1994

ISBN 0 19 577535 X

Printed in Pakistan at
Mueid Packages, Karachi.
Published by
Oxford University Press
5-Bangalore Town, Sharae Faisal
P.O. Box 13033, Karachi-75350, Pakistan.

Dedication

This book is dedicated to
all the Afghan men, women and children
who laid down their lives for
the honour and independence of Afghanistan.

Contents

Preface

My interest in writing this book was aroused by the tragic events in Afghanistan. Although the situation created by the Soviet intervention in Afghanistan attracted world-wide scholarly attention, little effort was devoted to understanding the domestic roots of the confrontations and explaining the linkages between internal strife and rivalry among the foreign powers. Most of the studies have focused on the Soviets' motives, the nature of the resistance, and the security climate of the region. This is not to discount the value of these works, but most of them suffer from two serious flaws. First, they represent, more or less, a situational analysis. Works by journalists reporting the events fall into this category. Scholars of Soviet studies, particularly those focusing on Moscow's Third World policies, also jumped on the Afghan scholarship bandwagon. They attempted to understand and explain Soviet behaviour by analogy, and more in the context of some assumed, and some real, expansionist aims of the Soviet Union. They tended to overlook the peculiarities of the Afghan situation; needless to say they had only a rudimentary knowledge of Afghan society. Anthropologists, who had studied Afghanistan for a long time, provided greater insights into the development of the resistance, the weak foundations of the Afghan state, and the patterns of local authority, but their analyses hardly touched the vital issues of political power, civil strife, regional and international security. Second, analytical frameworks and conceptual schemes embedded in cold war polemics have generated some heat but shed little light on the Afghan conflict.

In our view, two things would be necessary to advance scholarship in this field. First, we must combine insights from various social science disciplines in order to understand the social dynamics of the Afghan conflict. This particular aspect has been ignored due to the salience given to external intervention. Second, we need to explore linkages between the domestic, regional, and international environments of Afghanistan. An attempt has been made in this book to examine all the factors that account for the Afghan tragedy. We look at the nature of the Afghan state and society, the dynamics of regional and global power structures, the externalization of civil strife, and the resultant fragmentation of political power in the country. We hope that our perspective will contribute to the quality of debate on the politics and security of Afghanistan.

I have the pleasant responsibility of acknowledging my debt of gratitude to several individuals and institutions who aided the development of this work. Although this study has profited from a number of scholars, I am particularly appreciative of Richard S Newell, who read the entire manuscript and made valuable suggestions for its improvement. A number of friends and colleagues at the Quaid-i-Azam University took a deep interest in my work. I would specially like to mention M Rafique Afzal, Rais Ahmad Khan, Mohammad Waseem , Tahir Amin, Sikandar Hayat and Saeed Shafqat who gave freely of their time, advice and critical comments. I am very grateful to all those Afghans who took the time to talk to me at length about the political developments in their country. Space does not permit me to list all of them. It gives me special pleasure to thank Sayyid Ahmad Gailani, Pacha Gul Wafadar, Dr Faizullah Kakar, Dr Najibullah Mujaddedi, Engineer Abdul Rahim, and Professors Hasan Kakar and Anwar ul-Haq Ahady.

I wish to express my deepest appreciation to all the institutions that have at various stages supported my research and writing on Afghanistan. My gratitude goes to the Rockefeller Foundation for granting me an International Relations Fellowship, which enabled me to be based at the Institute of International Studies, University of California, Berkeley, for the academic year 1984/85. I gained much not only from the rich library resources but also from the very stimulating intellectual and social environment. I have benefited from exchanges with Professor Leo Rose, Agha Khalid Saeed and Fayyaz Baqir. I am grateful to the Hoover Institution, Stanford University, for awarding me a grant in the Fall of 1985. This allowed me access to the archives of the Institution and offered a great opportunity to discuss my research with the Stanford University faculty.

My greatest debt is to the Joint Committee on Soviet Studies of the American Council of Learned Societies and the Social Science Research Council for awarding me a fellowship in 1989 under the Program for the Development of Soviet Studies. I used this fellowship to attend the Russian Research Centre, Harvard University. It was in the congenial, scholarly atmosphere of Harvard that I was able to do most of the drafting of this book during 1990. I am profoundly grateful to Professor Adam Ulam, Eden Naby and James Critchlow, who were kind, generous and very supportive of my enterprise. I would like to express my gratitude to Ms. Rose di Benedetto who typed the initial script with speed and humour.

Since that time, the study has undergone extensive revisions and updating. I further refined and polished it during 1991-92 at the Southern Asian Institute, Columbia University after my appointment as Quaid-i-Azam Professor. At Columbia, I have greatly benefited from my discussions with Barnett R. Rubin, who was very generous in sharing his material on Afghanistan; I cannot thank him enough. Last, but not least, I also owe special thanks to Ameena Saiyid and Yasmin Qureshi of the Oxford University Press for bringing this work to the public.

It would have been impossible to complete this book without the understanding and support of my wife Khalida and children Faiza, Shan and Sana, who endured two long years of my absence from home.

Rasul Bakhsh Rais

Southern Asian Institute,
Columbia University, New York.
3 February 1993.

Chapter 1

Introduction

While the fall of the Marxist regime has ended the most tragic chapter of Afghan history, during which more than a million lost their lives and about five million became refugees, the future looks uncertain. The loose coalition of the Mujahideen parties that has replaced Najibullah has yet to develop a consensus on fundamental issues; nor has it acquired enough political capacity to undertake a genuine national reconciliation. The new leaders confront serious challenges to national unity in the shape of separate groups controlling different parts of the country along ethnic, tribal and sectarian lines. External military intervention, the structure and process of internal civil strife, and the politics of the Afghan resistance have contributed to the political fragmentation of Afghanistan. This is manifested in political polarization and a bitter struggle for power among the various Afghan Mujahideen parties. Their political conflict has the potential to be transformed into an ethnic conflict between the Pushtuns, fearing loss of power, and the coalition of Uzbek and Tajik groups from the north who seem to have gained greater political influence in Kabul, traditionally dominated by the Pushtun elite. Existing cleavages and the absence of democratic institutions may not only complicate the process of maintaining stable coalitions but also threaten an already fragile national unity.

Even before the Marxists captured power in April 1978 Afghanistan had never developed into an integrated political community in the modern sense, due to internal diversity and an adverse geopolitical environment; but it had traditional institutions and a political culture, based on the autonomy of tribes and other social groups, that allowed the state to organize its political power and exercise its 'restricted' authority.[1] The traditional political order in Afghanistan was disrupted in part by the dependent modernization of the state's coercive power,[2] but more by its lack of institutional capacity either to accommodate the rising political aspirations of a new middle class or to renegotiate the balance of political power with the traditional elites of civil society. Conflict between the state and society was sharpened further by the capturing of political power by the Moscow-inspired Afghan Marxists, and then by the extension of the Brezhnev doctrine by the former Soviet Union to stabilize its dependent elites in Kabul.

The central premises of this book are that a) the ruinous conflict in Afghanistan has significantly altered the balance of political and social forces in society without resolving traditional divisions; b) the political polarization and multi-level confrontations that evolved in the course of the civil war have further strengthened ethnic, tribal, territorial and religious identities; and c) given the transformation of Afghan society during the war and its fragmentation into several localized centres of power, Afghanistan now requires the structuring of a negotiated political legitimacy based on regional autonomy.

To understand the politics of revolution, war and resistance in the traditional Islamic society of Afghanistan that attracted a wide-ranging contest of regional and international power politics we must, therefore, combine insights from various social science disciplines, and explore interdependencies between the domestic and the international environments. In doing so, we must first look at the nature of the Afghan state and society, and the interactive processes between the two; second, we must examine the dynamics of regional and global power structures and their impact on the geopolitics of Afghanistan; third, we would have to discuss the externalization of civil strife in terms of the linkages between internal confrontations and rivalry among foreign powers. Finally, two questions need to be raised: a) how will change and continuity in the Afghan social structures affect the traditional patterns of authority and the future processes of structuring a new political order; and b) how will regional transformation and changes in global power relations after the cold war influence internal transition in Afghanistan.

The roots of the Afghan state never went deep enough in society to emerge as a focal point of the citizens' identity.[3] They remained underdeveloped both in terms of modern institutions, with the exception of the armed forces, as well as in their political capacity to extend direct and effective control over all subjects and territories. The central system of the state functioned on an oligarchical pattern through networks of patron-client relationships. Powerful and socially influential groups participated in the system to maintain their status. Informal power-sharing amongst elite groups, largely based on tradition, gave the political order stability and legitimacy.[4] Due to its exclusive nature, the system remained closed: recruitment to higher administrative positions and political roles was restricted to influential members of the patrimonial elite.

The 'old power' elites, satisfied with the distribution of power and autonomy in their local affairs, remained loyal to the system. Family ties among elites and their overlapping participation in tribal, religious and social networks further promoted their integration into the power arrangements at the top. The ruling establishment of Afghanistan, like any other in the Third World, had a stake in preserving the lopsided distribution of power and privileges in society, which came through ascription rather than personal achievement. In part it was this underdevelopment of civil society, made evident by the absence of any organized protest, that made possible the maintenance of a monopoly in exercising social and political control by the traditional institutions.[5]

Autonomous structures of local authority have long existed parallel to the state.[6] They have proved capable of blocking the state's access to the local population in the critical areas of political power and identity of the individual. Within this broad strand of polarization of state-based power and local authority, two factors inherent in the social and economic structures of Afghanistan have frustrated attempts to integrate the state with the fragmented sections of society. The first factor relates to the organization of social and political power in the vast and autonomous periphery of the state. The *khans, maliks, mullahs,* landlords and tribal chiefs have exercised far greater influence than the state Blood ties, kinship, tribal ties and their hold on local economies have contributed immensely to the power of the non-state societal elites. Local patron-client relationships and the exercise of authority, though within the bounds of tradition, have been socially accepted and generally perceived to be benevolent compared to an alien and intrusive image of the state.[7]

Second, the political economy of agricultural production constrained the growth of the state. Since arable land and water resources were limited and scattered, agriculture remained largely subsistence-oriented.[8] Production was localized, and just enough to sustain the local population, so economic transactions or trade across local boundaries did not develop.[9] Consequently, a national market system which could integrate the subsistence-oriented rural economies did not take off. The underdeveloped infrastructure further added to the fragmentation of the economy. With the construction of a few highways and the introduction of mechanized cultivation on a limited scale, agriculture began to become commercialized in the 1960s and 1970s, but changes in the agricultural economy were restricted to a few areas and its benefits confined to influential landowners who had easy access to government-sponsored loans.[10] Modern sectors of the economy, such as industry, commerce and finance,did not register any growth either. The important point is that a political economy based on agricultural production could not generate enough resources for the state to strengthen its authority.[11]

The traditional political order of the 'rentier' state,[12] though based on a somewhat elitist consensus, could not go on unchallenged. The selective modernization of the state apparatus through foreign resources brought forth a new class of administrators and intellectuals. This segment of Afghan society, though small, strongly influenced the political outlook of volatile young groups which socialized in the relatively modern atmosphere of the school. Social mobility and the prestige which administrative and professional positions conferred served as catalysts to the expansion and influence of modern educational institutions.[13]

The ideological orientation and the ethos of the groups that emerged from the modern school system clashed directly with the interests of the 'old power' elite. Their aspirations for representation and a share in power perforce challenged the traditional relations of authority. The attendant desire for political restructuring introduced a new intellectual discourse on competing ideas and ideologies.

Debates and controversies seeking a replacement of the authoritarian system of monarchy focused primarily on socialism, democracy and Islam.[14]

These conflicting ideologies of aspiring political groups both within and outside the country explain, to a great extent, the origin of the multipolar confrontations that have ravaged Afghanistan for the past many years. Two broad categories are the Marxism of the *Parcham* (banner) and *Khalq* (people) factions of the former People's Democratic Party of Afghanistan (PDPA), and the Islamic programmes of the radical Mujahideen groups.[15] Their ideological frameworks were not only distinct in terms of claims to political power but also represented contradictory visions of Afghanistan's political and economic restructuring. But these opposite ideological camps were never homogeneous entities even within their own ranks—their inner factionalism bordering on fratricide needs to be explained. First is the issue of interpretation of Islam as a state ideology which, beyond rhetoric, lacks consensus among the Islamic community. Islamist Mujahideen groups are no exception.[16] Nor have Marxist groups anywhere, let alone in the political milieu of Afghanistan, formed stable coalitions.[17] Class background, the socialization process of leaders and the absence of a democratic, pluralistic political culture are some of the variables that help us explain confrontations between the various ideological groups. The social and cultural orientations of the members of these groups were incompatible with the political values that would allow accommodation with those opposed to their beliefs.[18] Not surprisingly, they became poised to confront each other even in their preliminary stages of formation.

These contradictions and hostilities were further heightened by the introduction of the electoral process under what was by Afghan standards the liberal Constitution of 1964. Clear lines were drawn between the so-called Modernists and the Islamists, who had their genesis in the polarization of the 1920s, generated by the modernization programme of King Amanullah.[19] It is necessary to state from the outset that neither of these camps was politically broad-based: they functioned more in the urban environment, and mostly among the politically aware and literate sections of society. In each group there existed traditionalists and moderates, of both the Islamic and the secular variety, who favoured a more gradualist approach to the development and reform of traditional institutions.

This was the familiar pattern of ideological discourse witnessed in most post-colonial Islamic societies. The modernist camp was composed of elements from nationalist, secularist, and Marxist groups. Although the motivations and social origins of these groups were different, their political interests converged when confronting the rising power of the Islamist groups, which were equally opposed by the conservative religious establishment for what it called their *jadeed* (modern) interpretation of Islam.[20] It was, rather, a grand coalition of diverse interests that developed more expediently, but was not adequately structured.

While modernization, development and social reforms appeared to be the common political agenda of both groups, inherent social and class contradictions remained unresolved. Each group waited for the opportune moment to outflank the other. This became apparent when the coalition seized power in 1973 under the leadership of Sardar Daud, a former premier and member of the royal family.[21] He was a convenient choice of the modernist coalition in that he had earned the image of a true nationalist, and was credited with the expansion and modernization of the Afghan armed forces through Soviet assistance.

The Marxist groups that had existed independently were definitely encouraged by the growing partnership of the Soviet Union with their country, which was initiated by Daud during his premiership (1953-63). In the wake of increasing Soviet interest, the activities of the left wing groups were no longer confined to barren intellectual debates in drawing rooms. They found audience in educational institutions and among Afghan army personnel, who were trained in the Soviet Union and were quite receptive to their political propaganda. The revolutionary strategy of the Marxist groups, from the very beginning, stressed a sort of united front of all the progressive sections of Afghan society.[22] Recruiting and indoctrinating a cadre within the Afghan armed forces was central to their political campaign; otherwise, obstacles to a mass revolutionary line abounded. The society at large was religious, conservative and traditional in its values. Foreign influence, let alone the Soviet brand of socialism, generally perceived as atheistic and anti-Islamic, was an affront to the cultural sensibilities of a largely peasant society.[23]

Another factor inhibiting the adoption of a grassroots revolutionary movement was the fact that the Marxist groups lacked a unified and broad-based political organization. Nor did their actions reflect any realistic thinking on the relationship between the masses and the revolution. Furthermore, they had sharp divisions among themselves on ideological issues and political tactics. Personality clashes and serious differences over the leadership of their movement kept the Marxists divided and quite antagonistic towards one another.[24]

The temporary unity which political events in the country dictated at the time of the revolution disappeared sooner than was expected.[25] While in power, the new socialist elite demonstrated a serious lack of pragmatism. Besides, a sort of text-book revolutionary thinking and Marxist romanticism, rather than a practical approach to restructuring the state and cultivating links with the traditional society, guided their actions. Their political ethos, expressed in rhetorical terms, and an obsession with creating new social relations, put the Marxists on a collision course with the masses in society.[26] The collision occurred before they had achieved complete control over all the institutions. In fact, they had not even managed to develop any political consensus among themselves. Conversely, the leaders of the PDPA thought that by dominating the coercive institutions at the top, they would effectively suppress opposition and resistance from what they frequently referred to as the 'vested interests' and 'reactionary' elements. Attempts to subordinate the state and society to a Leninist party through coercive means was, indeed, the repetition of an old practice, which backfired in the Islamic social and cultural environment of Afghanistan.

Two factors, largely missing in the political analysis of the Marxist regime, added to its difficulties. First was the fact that the revolution was primarily a political change at the top through a military *coup d' etat*. Even within the armed forces, there was hardly any consensus on the meaning and direction of the revolution. The so-called 'revolutionary legitimacy' quite often cited in the case of authoritarian regimes in the Third World was not to win the hearts and minds of the Afghan population.[27] If social structures are important, as they are in the acceptance and stability of the power elite, then the Marxist rulers of Afghanistan were definitely outsiders. They had neither social claims to power nor any basis for making a convincing political case for being where they were. In the context of Afghan political culture they were non-entities. Military leaders in many Third World countries crashed through the doors of political power but many of them spoke in a political idiom which was compatible with common values, or at least, not too openly offensive to popular beliefs. In Afghanistan, popular perceptions of the revolution led by the Marxists, their ideology and their links with Moscow were not conducive to even preliminary political communication with the masses.

Second, it is now apparant that the heavy reliance on the application of state power against the rebellious masses was counterproductive.[28] Many of the influential elements of the Afghan Marxist power structure were Stalinist in their approach to controlling dissent. Instead of initiating a meaningful dialogue in vernacular political vocabulary, they strengthened the intelligence agencies, established torture cells, filled the prisons beyond capacity, and resorted to general warfare against the common man. This was not an appropriate strategy for achieving the transformation of the country, nor was it capable of offering a basis for integrating civil society into the new political order.

The administrative apparatus of the state, which was never strong, began to disintegrate in the wake of mass uprisings. The revolt grew stronger with each violent action of the Khalq functionaries. Popular resistance to the regime, led by socially influential sections of society, posed a serious challenge to the survival of the Marxist state. Fearing its collapse, the Soviet Union substantially increased its level of support. It poured in more weapons and stationed more advisers to stabilize the shaky revolution and protect the insecure Afghan Marxists. This open collaboration and political identification with the Soviet Union further removed the Afghan Marxists from the mass of society.

Some attitudinal factors rooted in Afghan culture also served to widen the gulf between the oppressive Marxist state and civil society.

Conflict between the two generated a vicious cycle of action and reaction in the the conduct of violence. *Badal* (revenge) and *dushmani* (animosity) have long fired social and tribal feuds. Both Islam and Afghan tradition attach high personal value and prestige to these concepts. More important is what Nazif Shahrani calls the culture of *Jihad* (holy war), which has mobilized the Afghans and highly motivated them to take up arms against atheistic Marxists.[29]

The second element in our analytical scheme is the geopolitical environment of Afghanistan. There is no doubt that internal conflicts, depending on their severity and the identification of the rival parties with extra-national forces, attract outside powers. But externalization of civil strife cannot be explained by the internal political process alone, more so in the case of Afghanistan. Historically, the country played an important role as a buffer between British India and the Czarist empire.[30] Britain for long exercised influence over the external relations of Afghanistan (1879-1919) with the objective of preventing Kabul from developing any strategic linkages with the former Russian Empire. The departure of Britain from the region in 1947 and its decline as a global power altered the traditional geopolitical balance in the region. Kabul's disputes with the new state of Pakistan over the Durand Line and Pushtunistan[31] issues added to its urgency to look for political and material support from the Soviet Union. The United States, the post-war superpower, preferred a security alliance with Pakistan rather than Kabul for its containment strategy.[32] Though Washington provided considerable development assistance to Afghanistan, it did not match the growing influence of Moscow in the important areas of training and supply of military equipment.[33] These were vital instruments to the Soviet Union's penetration of the Afghan state.

Afghanistan's landlocked character, its large and powerful neighbours and its undefined and undefended borders present another set of factors that opened the country to external interventions. Equally important to consider are the strategic goals, political interests, and ideological imperatives of the neighbouring states. Soviet intervention shaped by these concerns invited counter-intervention from Iran, Pakistan, Islamic states of the Middle East, and the Western powers. Superpower rivalry in a climate of cold war introduced a larger mix of external influences.

The Soviet military presence across the Khyber Pass changed the buffer status of Afghanistan, complicating the security dilemma for Pakistan which faced a vastly powerful India along its eastern flank.[34] The inflow of refugees and Mujahideen further dragged Pakistan into accepting a 'front-line' state role. But this policy decision, though influenced by geopolitical considerations, was not entirely independent of the domestic political process inside Pakistan. A military regime which was almost totally alienated found a splendid opportunity in the Soviet invasion of Afghanistan to attract western support.[35] The United States interpreted the Soviet move as a threat to the stability of the adjacent areas that are vital to the economies of the industrialized states. To safeguard the collective interests of the western world, Washington declared that 'any attempt by an outside force to gain control of the Persian Gulf region will be repelled by any means, including military force.'[36] Therefore, the conflict in Afghanistan, particularly following the Soviet invasion, was internationalized. Afghanistan was, from now on, no less a victim of its geopolitical factors than of its internal confrontations.

The linkages and interconnectedness between Afghanistan's internal political dynamics and its geopolitical environment need to be understood. Three factors—dependence of the state and the counter-elite on external factors, convergence of political and strategic interests of the Afghan partisans and their foreign supporters, and involvement of transnational ideological groups—help explain this interplay.

The dependence of the Afghan state upon the Soviet Union in modernizing its armed forces and launching development projects increased its vulnerability to the big neighbour's penetration in important areas of public policy and national security.[37] Moscow's involvement in the political affairs of Afghanistan further deepened with the Saur revolution. Lacking domestic support, the Afghan Marxists leaned heavily on the Soviet Union to ensure their political survival. Without Soviet commitment and direct participation, the fragile political order which they were attempting to build could not stand in the face of a highly mobilized nation-wide resistance. Equally important was the convergence of Soviet regional interests with a dependent Marxist regime in Afghanistan. These were: to counter the rising power of the Islamic movements in the region, which had the potential of influencing political developments in the Soviet Union's Central Asian republics; to ensure stability in the bordering areas; and to demonstrate the ability to intervene militarily to defend its allies in the region. However, any explanation of Soviet intervention has, to date, remained controversial. In our view the more immediate interests of the Soviet Union which guided its decision to intervene related to the internal political dynamics of Afghanistan. The Soviets were convinced that Hafizullah Amin lacked the credibility to promote reconciliation among the warring factions of the PDPA, a conflict that had been exacerbated by the murder of Nur Mohammad Taraki, the founding father of the Marxist movement.[38] More important, the Soviets did not fear any counter-move by the United States as Washington's security arrangements in south-west Asia had collapsed with the success of the Islamic revolution in Iran.[39] The Soviet leaders, of whom only a very limited number participated in the decision to intervene in Afghanistan, assumed that military action could salvage the political situation in Afghanistan, which was fast slipping into anarchy.

The initial goals of the military intervention were: a) to remove the untrustworthy Amin, b) to install the more pliable Babrak Karmal, and c) to pacify the country. While assigning these tasks to their armed forces, the Soviet planners failed to take into account the objective political conditions of Afghanistan. Encouraged by their previous successes in similar operations in Eastern Europe,[40] the Soviet leaders pinned high hopes on their military organization, state-of-the-art equipment, and a counterinsurgency doctrine based on the massive and indiscriminate application of force, not realizing that the combat environment of Afghanistan was vastly different from anything they had experienced earlier. The physical conditions of the country and the intensity of the resistance put up by the population frustrated Soviet attempts to restore law and order and to establish an effective government. The presence of the Soviet forces was offensive to the national sentiment of the Afghans and thus counterproductive in resolving the crisis. Inviting the Soviets to intervene to help them secure their political survival further stigmatized the PDPA leaders, alienating them from society.

The use of massive force against the civilian population, the destruction of the rural infrastructure and the ruthless enforcement of the 'scorched earth' strategy strengthened political support for the resistance which took up arms against the Kabul regime. The more the Soviets escalated their level of counterinsurgency operations, the more they moved the warring Afghan groups away from national reconciliation. Thus Soviet involvement in support of the Kabul regime was politically divisive and a failure as far as securing its objective went, i.e., pacification of the country. In this respect, this study focuses on the following questions: Was the political climate of the country amenable to the fostering of an effective military role by a foreign power? Was the structure and strength of the forces adequate to crush the Afghan resistance? Can a modern army organized, trained, and equipped to fight a high-intensity war win against a ferocious, broad-based, and popular resistance?

The defiance put up by the Afghan resistance forced the Soviet Union to fight a war of attrition which did not work to its advantage. The war took a heavy toll in men and material and seriously damaged the Soviet Union's international reputation. After utterly failing to expand control beyond the cities despite protracted counter-insurgency campaigns, the Soviet Union decided to seek a way out of the Afghan quagmire.[41] Political changes within the Soviet Union, especially the new leadership committing itself to openness and restructuring, added to the previous urgency to end a war that was unpopular at home and had generated tremendous opposition abroad, particularly from the West.

Among all the factors that contributed to the decision of the Soviet Union to leave Afghanistan, the spirited and intense guerrilla operations of the resistance were central to its calculation. The Mujahideen succeeded to a considerable degree in achieving the fundamental objectives of their campaign: a refusal to allow the consolidation of the Kabul-Soviet forces, the denial of legitimacy to the Kabul regime and its general alienation from the masses. The ensuing stalemate hurt Soviet morale and its staying power more than it affected the determination of the resistance groups to liberate their country.

The Afghan resistance movement itself constitutes a fascinating study of modern guerrilla wars and national liberation movements. Was it a traditional tribal revolt? Why has the Afghan resistance remained fragmented? What is the nature of the differences between the resistance groups? How would these influence the future politics of the country? What factors stalled their military victory? These are some of the questions which have been raised in this study. Social structures and general organizational patterns of society must be taken into account in order to understand the evolution of wars of resistance and their operational styles. Unlike many other movements, the Afghan resistance was never united, even in the face of a formidable enemy and the commonality of objectives. This was a reflection of ethnic, linguistic, religious and regional differences in Afghan society.[42]

Afghan tribes, ethnic groups and village chiefs have jealously guarded their traditional autonomy over the centuries. When it comes to leadership issues or exercising influence within their areas of jurisdiction, they submit solely to the locally established authority structures.[43] Emerging in the wake of a national uprising, the resistance groups were organized, first and foremost, on a local basis. However, the growing need for military equipment, training and general support forced them to enter into loose alliances with the political parties set up by more prominent figures who served as a conduit for foreign aid.[44]

As the conflict progressed, and the Soviet hold over the country looked increasingly unstable, foreign powers interested in rolling back the Soviet tide penetrated the political structure of the resistance movement through money, weapons and political support. Their influences penetrated deep into the Afghan society in exile, particularly amongst Afghan parties operating from sanctuaries in neighbouring countries. With the converging of political and strategic interests between the rival Afghan parties and the external powers, two parallel alignments of domestic and foreign actors developed. The Moscow-Kabul combination produced counter-coalitions of the Mujahideen with Iran, Pakistan, various Islamic countries and the United States, gathering general consensus and the support of the industrialized nations. These powers shared the common objective of forcing the Soviet Union out of Afghanistan and installing a representative political authority in Kabul. As we shall examine in some detail, beyond this minimal consensus the coalition partners had quite divergent interests and motivations. With the withdrawal of the Soviet forces, their contradictions became more open.

The role of transnational ideological groups and movements in subjecting Afghanistan to external intrusion is no less significant than that of the state actors. Not only did they mobilize political support in the Islamic countries and generate financial resources, but they also trained and sent their volunteers to fight along side the Afghan Mujahideen groups.[45] The waging of *jihad* by these private transnational groups points to a new phenomenon of the Islamic revivalist movement. But this foreign factor has added to the complexity of the intra-Mujahideen conflict, as members of the non-Afghan trans-Islamic groups have tended to offend popular religion, showing no regard for local culture. Driven by religious and ideological zeal, the operations of these groups are likely to influence domestic conflict in other Islamic societies where similar confrontations might shape up on the Afghan pattern.

Since the Soviet withdrawal, fundamental changes have occurred in the internal and external environment of Afghanistan. Internally, the resistance movement failed to formulate a joint military strategy to capture the urban areas or to expand its territorial control beyond traditional strongholds. Personal rivalries, infighting, ideological divisions and lack of political consensus stifled all moves toward forming a unified command structure. The differences between the Pushtun majority groups and the minorities on the one hand, and the political rift between the parties professing traditional and revolutionary Islam on the other, widened the conflict within the resistance, and with the departure of the common enemy, the Soviet troops, from the scene this rift became even more evident.

The Mujahideen's military debacle around Jalalabad, close to Pakistan's borders, in early 1989, coupled with the failure of dissident elements within the ruling Marxist groups to topple the Government from within, demonstrated that political change in Kabul could not be brought about by military means or intrigues alone. The Marxist regime, which had begun projecting an altogether new image of itself since 1986, moved quickly to refashion its ideology.[46] The collapse of communism in Eastern Europe, and the movement within the Soviet Union toward a market economy and openness, had a profound effect on the politics of client Marxist elites in the Third World. Largely under the impact of this global transformation, the Kabul regime shifted its ideological position away from the Soviet-inspired socialist model of modernization to more traditional ways of politics.[47] But attempts to rectify past wrongs came too late, and the moves to seek reconciliation were blocked by the perceptual gap between the regime and society.

Two issues continued to frustrate Kabul's efforts to broaden its support base. First was its legacy of repression. The deep scars caused by the collaboration of the Marxist groups with the Soviet interventionist forces were not likely to disappear soon. The scale of devastation and human suffering was so colossal that it presented strong psychological barriers against the regime. It is, equally, important to understand the impact of the Afghan tradition of animosity, which still exerts a strong influence on Afghan society.

Second, the regime, even with the best of intentions, lacked credibility in the eyes of its armed opposition. The resistance groups contended that the regime neither changed its character nor could be acceptable under any guise. Many of the initiatives of the regime toward national reconciliation, such as its pledge to hold free elections, and even its willingness to form a national government which could include the resistance groups, were therefore smashed upon the rocks of distrust. The antagonism between some of the Islamic resistance groups and the Marxist leaders is as old as the first initiatives of the former Soviet Union to penetrate Afghan society. The Soviet war and the civil strife played an exacerbating role in this historical divide. However, war-weariness on both sides, stalemate in the conflict, and fundamental changes in regional geopolitics and in the international power structure, seem to have generated considerable support for a political settlement among a majority of the Afghan groups, including some elements of the now defunct Marxist regime.

The most important change in the international environment of Afghanistan was the disintegration of the Soviet Union, which had played a pivotal role in defending the Marxist regime in Kabul. The Central Asian states, some of which now share borders and ethnic groups with Afghanistan, have attained sovereign status. The changes in the internal structure and international personality of the former Soviet Union have transformed its traditional rivalry with the United States into a growing partnership.[48] This was partly demonstrated in their agreement to halt military assistance to the warring Afghan parties from 1 January, 1992.[49] Once the supplies of arms from Moscow stopped, and the economic crisis within the former Soviet Union made it practically impossible to meet previous commitments of food and fuel to Kabul, the regime began to crumble from within.[50] It is interesting that without foreign support the Marxist regime of Najibullah did not survive beyond 115 days.[51]

Washington also redefined its security links with Pakistan, which had played a central role in the Afghan resistance. No longer important to US regional strategy, Pakistan's economic and military assistance programme has been terminated, though under a public law which requires the President of the US to certify at the beginning of each fiscal year that Pakistan does not 'possess a nuclear explosive device.'[52] Reassessing its own regional role in the wake of the independence of Central Asian states, Islamabad shifted its policy from wanting a total military victory by the Mujahideen to supporting the United Nations plan for a political settlement.[53] The credibility and influence of this world body, long hamstrung by cold war power politics, has significantly increased in the changing global relations. It is a testimony to this fact that never before have UN peace initiatives on Afghanistan received as much consensus and support—from regional powers, moderate Afghan guerrilla groups, and the world powers—as they did before the sudden collapse of the Marxist regime in April 1992. This was, indeed, a sign of positive change towards resolving one of the bloodiest conflicts of the cold war era but, unfortunately, the internal struggle for power among the Mujahideen groups overtook the UN-mediated political settlement, which had offered the best hope of paving the way toward a peaceful transition.

Although some factions of the Mujahideen parties have taken control of Kabul and the provincial towns, restoration of peace in

Afghanistan will depend on how the issue of political power is settled and what the mechanism for restructuring political legitimacy will be. The only political solution of the Afghan conflict which can hold is one consistent with Afghan social structures and historical experiences. Therefore, initiatives aimed at domestic reconciliation will not produce results until they seek the sufficient involvement of local political authority and rely on traditional institutions, such as the *jirga*. But what is critical is that the diverse collection of internal and external actors directly or indirectly involved in the Afghan conflict must avoid forcing upon the Afghans some new form of political authority or settlement from above. A political solution to the conflict, aimed at reconstituting broad and legitimate power arrangements, would be credible only if it gathers the support of all the important Afghan groups. As a first step, outside powers that have exploited the geopolitical situation in Afghanistan must completely and unequivocally disengage themselves. No matter how well-meaning it appears, the involvement of these powers would undoubtedly complicate the task of structuring national consensus among the Afghans, because foreign benefactors and friends are a poor substitute for indigenous political or social forces in mediating the conflict among the old and new power elites. It is essential to recognize that peace and stability in the fractured polity of Afghanistan might not be restored unless those engaged in the struggle for power, or outsiders wishing to defuse it, seek new relationships among all the constituent groups—ethnic, regional and religious. Any attempt to structure a powerful centralized authority within the framework of the nation-state model would be self-defeating. Decentralization, regional autonomy and revitalization of traditional patterns of authority would all strengthen accommodation. Political structuring under catchy slogans of constitutionalism, electoral process, and even self-determination, would lose significance if it ignored the underlying social forces of Afghan political culture.

NOTES

1 *Jirga* (grand assembly of elders and notable) and provincial councils of *sardars* (chiefs) were two traditional institutions that, along with the religious establishment, played an important role in conferring legitimacy on the Afghan kings, and approved constitutional and legal changes in the country. The institution of *jirga* also functioned at local levels to adjudicate disputes between individuals and tribes. Looking at the evolution of the Afghan state, particularly its early phase, and its central system of governance, one can hardly miss the point that it rested on the consent of the tribes and other socially influential groups. Allegiance to the state by them was conditioned by the unwritten but well-respected tradition of local autonomy. On this point, *see,* A Olesen, 'Afghanistan: the Development of the Modern State,' in K Ferdinand and M Mozaffari (eds), *Islam: State and Society,* (London: Curzon, 1988), pp. 155-169; Jolanta Sierakowska-Dyndo, 'The State in Afghanistan's Political and Economic System on the Eve of the April 1978 Coup,' *Central Asian Survey,* Vol. 9, No. 4, 1990, pp. 85-86. Leon B Poullada has argued that 'originally the central government was in effect an emanation of a tribal confederation and the Amir was considered a paramount chief, a *primus inter pares,* by other tribal chiefs'. Other factors that contributed to the autonomy of the tribes are: geographic location, tribal culture of defiance, dynastic quarrels and the impact of the Great Game between Britain and Russia, which, in particular, promoted the pugnacity of the Pushtun tribes.' *See,* Poullada, 'Political Modernization in Afghanistan: The Amanullah Reforms', in George Grassmuck *et al, Afghanistan: Some New Approaches,* (Ann Arbor: The University of Michigan, 1969), pp. 116-17.

2 Modernization of the military was the fundamental tool, since early this century, by which the Afghan monarchs attempted to consolidate the power and authority of the central state and preserve internal political order. This was also meant to lessen the dependence of the state on the tribes for raising forces to quell rebellions. *See,* Patrick J Reardon, 'Modernization and Reform: The Contemporary Endeavor,' in Grassmuck, op. cit., p. 151.

3 On the evolution and identity of the Afghan state and its relationship with the civil society, *see,* Barnett R Rubin, 'Lineage of the State in Afghanistan,' *Asian Survey,* November 1988, pp. 1183-1209; M. Nazif Shahrani, 'State Building and Social Fragmentation in Afghanistan: A Historical Perspective,' in Ali Banuazizi and Myron Weiner, (eds), *The State, Religion, and Ethnic Politics: Afghanistan, Iran and Pakistan,* (Syracuse, N.Y. 1986), pp. 23-74.

4 Three types of leadership may be identified which used links with the state to enhance their local power: elders of tribes and ethnic groups, *ulema* (scholars of Islam), and *rohanyun* (saintly figures), which included *pirs* (spiritual leaders) and *hazrats* (respectable on religious grounds). *See,* Nabi Misdaq, 'Traditional Leadership in Afghan Society and the Issue of National Unity,' *Central Asian Survey,* Vol. 9, No. 4, 1990, pp. 109-12.

5 Although social class distinctions existed in Afghanistan, the basic identity of the common man in the villages was with a clan or tribe. The mystique of kinship relations prevented the emergence of any challenge to traditional authority.

6 On the nature of the political system of Afghanistan and its social and historical roots, *see* a pioneering study by Richard S Newell, *The Politics of Afghanistan,* (Ithaca: Cornell University Press, 1972).

7 For this analysis I have benefited from David Gibbs, 'The Peasant as Counter-Revolutionary: The Rural Origins of the Afghan Insurgency,' *Studies in Comparative International Development,* Vol. 21, No. 1, Spring 1986, pp. 37-45.

8 Estimates of arable land vary, but it is generally believed that land area where something can be grown is about 12 per cent, of which only 4 per cent is cultivated, mainly due to insufficient irrigation system. Louis Dupree, cited by Gibbs, in *Ibid.,* p. 38.

9 Maxwell J Fry, *The Afghan Economy: Money, Finance, and the Critical Constraints to Economic Development,* (Leiden: E J Brill, 1974), p. 56.

10 Douglas G Norvell, *Agricultural Credit in Afghanistan: A Review of Progress and Problems from 1954 Until 1972,* (Kabul: Agricultural Division, United States Agency for International Development, 1972), pp. iii, 1.

11 Despite the largest share of agriculture in Afghanistan's economy, the government's farm-generated revenues were nominal. It is difficult to find statistics on this issue. Just to give an idea, government revenues in 1971-2 were only 1.2 per cent. K Glaubitt, Saadeddin and B Schafer, 'Government Revenues and Economic Development of Afghanistan,' *Afghanistan Journal,* Vol. 4, No. 1, 1977, p. 1.

12 Barnett R Rubin applies 'rentier' state model to Afghanistan by suggesting that, since 1957, more than 40 per cent of the state's income came from foreign sources. *See,* his 'Political Elite in Afghanistan: Rentier State Building, Rentier State Wrecking,' *International Journal of Middle East Studies,* No. 24, 1992, p. 78.

13 On the expansion and prestige of modern education, *see:* Wolfram Eberhard, 'Afghanistan's Young Elites,' *Asian Survey,* February 1962, pp. 3-22; Barnett R Rubin, 'The Old Regime in Afghanistan: Recruitment and Training of a State Elite,' *Central Asian Survey,* Vol. 10, No. 3, 1991, pp. 73-100; Olivier Roy, *Islam and Resistance in Afghanistan,* (Cambridge: Cambridge University Press, 1986).

14 Tahir Amin, *Afghanistan Crisis: Implications and Options for Muslim World, Iran and Pakistan,* (Islamabad: Institute of Policy Studies, 1982), pp. 95-106.

15 On the education, socialization and development of ideological identities of the Marxist and Islamist elites, *see,* Rubin, 'Political elites in Afghanistan: Rentier State Building and Rentier State Wrecking,' op. cit., pp. 78-99.

16 On political and ideological differences, *see,* Shah M Tarzi, 'Politics of the Afghan Resistance Movement: Cleavages, Disunity, and Fragmentation,' *Asian Survey,* Vol. 31, No. 6, June 1991, pp. 479-511.

17 Raja Anwar has provided the most detailed and insightful analysis of this subject. *See,* his *Tragedy of Afghanistan: A Firsthand Account,* translated from Urdu by Khalid Hasan (London: Verso, 1988).

18 Both the Afghan Marxists and Islamist ideologues and political activists came largely from tribal and feudal social backgrounds. This is why they did not show any tolerance to each other. For Islamists, all Marxists were *kafirs* (atheists, non-believers) and agents of the Soviet Union, while Marxists regarded the Islamists as a fundamental obstacle to their rule and reform programme.

19 Efforts of the Afghan state to modernize the army and civilian bureaucracy began in the late 19th century (reign of Amir Abdur Rahman, 1880-1901), but it was King Amanullah who provided greater political will and commitment. *See,* Leon B Poullada, *Reform and Rebellion in Afghanistan, 1919-1929: King Amanullah's Failure to Modernize a Tribal Society,* (Ithaca: Cornell University Press, 1973).

20 Amin, *Afghanistan Crisis,* pp. 95-101.

21 For an analysis of the overthrow of the monarchy in 1973, *see,* Louis Dupree, 'A Note on Afghanistan 1974,' *American University Field Staff Reports* (AUFS), South Asia Series, Vol. 18, No. 6, 1974.

22 *See,* the founding document of PDPA in Anthony Arnolds, *Afghanistan's Two Party Communism,* (Stanford: Hoover Institution Press, 1983), pp. 137-48.

23 M Nazif Shahrani, 'Introduction: Marxist "Revolution" and Islamic Resistance in Afghanistan,' in M N Shahrani and Robert L Canfield (eds.), *Revolutions and Rebellions in Afghanistan,* (Berkeley: Institute of International Studies, University of California, Berkeley, 1984), pp. 3-57.

24 For a detailed examination, *see,* Anwar, op. cit.

25 The leaders of Parcham were ousted from power within the first three months of the revolution. First, Babrak Karmal, who had been appointed deputy prime minister, was sent abroad as ambassador in July 1978, and then in November was dismissed from this position. Other prominent members of his party were also removed from their government and party positions and persecuted.

26 Imposition of alien revolutionary concepts on a largely feudal and tribal society without a proper understanding of its structures, values, and critical forces was, perhaps, the central cause of the revolt against the Marxist government.

27 William Maley has very convincingly demonstrated that the Leftist regime confronted a serious legitimacy crisis. *See,* his 'Political Legitimation in Contemporary Afghanistan,' *Asian Survey,* Vol. 27, No. 6, June 1987, pp. 705-25.

28 This can be gauged from the fact that in the first phase of the conflict (1978-9), before the Soviet invasion, 40,000 Afghans were killed in 1978 and 80,000 in 1979. The rate of casualties increased in the following years. It is estimated that between 1 million to 1.5 million might have been killed by the time Moscow pulled out its troops. The proportion of those incapacitated by the war is 31 per thousand of the entire population of the country. About one-third of the citizens became refugees in neighbouring countries, mostly (3.5 million) in Pakistan. *See,* a well-documented study by Marek Sliwinski, 'Afghanistan: The Decimation of a People,' *Orbis,* Vol. 33, No. 1, Winter 1989, pp. 39-55.

29 Shahrani, 'Introduction: Marxist 'Revolution' and Islamic Resistance in Afghanistan,' pp. 25-40; *See also,* Eden Naby, 'The Concept of Jihad in Opposition

to Communist Rule: Turkestan and Afghanistan,' Studies in *Comparative Communism,* Vol. 19, No. 3/4, Autumn/Winter 1986, pp. 287-300.

30 On Afghanistan's geopolitics and foreign relations, *see,* Ludwig W Adamec, *Afghanistan, 1900-1923: A Diplomatic History,* (Berkeley: University of California Press, 1967); L W Adamec, *Afghanistan's Foreign Affairs to the Mid-Twentieth Century: Relations with the USSR, Germany, and Britain,* (Tucson: University of Arizona Press, 1974); V Gregorian, *The Emergence of Modern Afghanistan: Politics of Reform and Modernization, 1880-1946,* (Stanford: Stanford University Press, 1969); Olaf Caroe, *Soviet Empire,* (London: Macmillan, 1954); William P Coates and K Zelda, *A History of Anglo-Soviet Relations,* (London: Lawrence and Wishart, 1943; Arnold Fletcher, *Afghanistan: Highway of Conquest* (Ithaca: Cornell University Press, 1965); David Fromkin, 'The Great Game in Asia,' *Foreign Affairs* Vol. 58, No. 4, Spring 1980, pp. 936-51; Zalmay Khalilzad, *The Return of the Great Game* (Santa Monica, Ca.: California Seminar on International Security and Foreign Policy, September 1980).

31 The Durand Line is the present boundary between Afghanistan and Pakistan. An accord to draw this boundary was signed between Britain and Afghanistan in 1893. After the creation of Pakistan, Afghanistan insisted on the redrawing of this boundary, which Pakistan has claimed to be final. Pushtunistan is the Afghan title for the tribal territories around the Durand Line that form part of Pakistan. Afghan rulers have from time to time demanded that those tribal areas inhabited by the Pushtu-speaking ethnic groups either join Afghanistan or become an independent Pushtunistan state. These two issues caused considerable friction between the two neighbouring states in the 1950s and 1960s. Afghanistan's position changed considerably in 1976-77 when Sardar Daud decided to improve relations with Iran and Pakistan. They lost significance with the civil war, the Soviet invasion and Pakistan's support for the liberation of Afghanistan. Whether or not the future rulers of Afghanistan will take up old quarrels with Islamabad is yet to be seen. On the historical origins of these disputes, *see,* S M Burke and Lawrence Ziring, *Pakistan's Foreign Policy: A Historical Analysis,* (Karachi: Oxford University Press, 1990), pp. 68-90; Ainslee T Embree (ed.), *Pakistan's Western Borderlands,* (Durham, N.C.: Carolina Academic Press, 1977), Mehrunisa Ali, *Pak-Afghan Discord: A Historical Perspective, Documents 1853-1979,* (Karachi: Pakistan Study Centre, Karachi University, 1990); Saeeduddin Ahmad Dar (ed.), *Selected Documents on Pakistan's Relations with Afghanistan 1947-85,* (Islamabad: National Institute of Pakistan Studies, Quaid-i-Azam University, 1986)

32 On this point *see,* Henery S Bradsher, *Afghanistan and the Soviet Union,* (Durham : Duke University Press, 1985).

33 In 1947, then again in 1955, the Afghan government had requested the United States for military aid. Kabul's disputes with Pakistan complicated the affair, but the United States supplied Afghanistan with $ 800 million in economic assistance between 1946 and 1978. The military assistance during this period amounted to only $5.6 million dollars. Milton Leitenberg, 'United States Foreign Policy and

the Soviet Invasion of Afghanistan,' *Arms Control,* Vol. 7, No. 3, December 1986, pp. 272-3.

34 Theodore L Eliot, Jr., and Robert L Pfaltzgraff, Jr., (eds.), *Red Army on Pakistan's Borders,* (Washington, D.C. : Pergamon-Brassey's, 1986); Francis Fukuyama, *The Security of Pakistan: A Trip Report,* Rand Report N-1584-RC, (Santa Monica: Rand Corporation, September, 1980); Lawrence Ziring, 'Soviet Policy on the Rim of Asia: Scenarios and Projections,' *Asian Affairs,* Vol. 9, No. 3, Jan-Feb., 1982, pp. 135-46; Robert G Wirsing, 'Pakistan and the War in Afghanistan,' *Asian Affairs, An American Review,* Vol. 14, No. 2, Summer 1987, pp. 57-75.

35 *See,* Omar Noman, *Pakistan: Political and Economic History Since 1947,* (London: Kegan Paul International, 1990), pp. 120-5; Shahid Javed Burki (ed.), *Pakistan Under the Military: Eleven years of Zia Ul-Haq,* (Boulder: Westview Press, 1991).

36 On the US response, *see,* : Richard P Cronin, *Afghanistan: Soviet Invasion and US Response,* IB 80006 (Washington, D.C. : Congressional Research Service, Library of Congress, updated 24 November 1981); US Congress, House Committee on Foreign Affairs, Subcommittee on Europe and the Middle East, *Hearing: East-West Relations, United States-Soviet Relations,* 24 and 30 January 1980, 96th Congress, 2nd Session (Washington, D.C.: US Government Printing Office, 1980); Zalmay Khalilzad, 'Afghanistan and the Crisis in American Foreign Policy,' *Survival,* Vol. 22, No. 4, July-August, 1980, pp. 151-60; William E Griffith, 'Superpower Relations after Afghanistan,' *Survival,* Vol. 22, No. 4, July-August, 1980, pp. 146-50; US Department of State, *Soviet Invasion of Afghanistan,* Special Report No. 70, April 1980.

37 Amin Saikal notes that 'by the turn of the 1970s no major Afghan political decision could be taken or military operation initiated without the knowledge of Soviet advisers and embassy,' Amin Saikal, 'Russia and Afghanistan; A Turning point?' *Asian Affairs,* Vol. 20, (old series 76), part II, June 1989, p. 173.

38 For an opposite view of Amin and the intrigue resulting in his murder, *see,* Beverley Male, *Revolutionary Afghanistan,* (London; Croom Helm, 1982).

39 On 8, 11, 15, 17, and 27 December 1979, the United States conveyed its serious concern to the Soviet Union about the movement of its troops around Afghanistan. But the capability of the US to prevent the impending Soviet invasion or retaliate effectively was not taken seriously by Moscow. *See, East-West Relations,* op. cit., pp. 111-21.

40 On these similarities, see Jiri Valenta, 'Prague to Kabul: The Soviet Style of Invasion,' *International Security,* Vol. 5, Fall 1980, pp. 114-41.

41 For an analysis of shifts in the Soviet approach to negotiations, *see,* an excellent study by Riaz M Khan, *Untying the Afghan Knot: Negotiating Soviet Withdrawal,* (Durham: Duke University Press, 1991), pp. 242-84.

42 These differences are discussed in detail in Chapter 4. *See also,* Grand M Farr and John G Merriam (eds.), *The Afghan Resistance; The Politics of Survival,* (Boulder: Westview Press, 1987); Ralph H Magnus, (ed.), *Afghan Alternatives,* (New Brunswick, N.J. : Transaction Books, 1985); Robert Canfield, 'Afghanistan: The

Trajectory of Internal Alignments,' *Middle East Journal,* Vol. 43, No. 4, Fall 1989, pp. 635-48.

43 Olivier Roy, 'Modern Political Culture and Traditional Resistance,' in Bo Huldt and Erland Jansson (eds.), *The Tragedy of Afghanistan: The Social, Cultural and Political Impact of the Soviet Invasion,* (London: Croom Helm, 1988), pp. 106-13.

44 Aid to the Afghan political parties was channelled through the Government of Pakistan, mainly by its Inter-Services Intelligence (ISI). The ISI distributed the weapons and organized the relevant support system. It used its own discretion as to which Afghan group would get what type of weapons and in what quantity. Author's interviews with Afghan resistance leaders in Peshawar, Pakistan, January 1990.

45 There is hardly any documentation of the numbers of foreign Islamist volunteers involved in fighting alongside the Mujahideen. According to investigations by this author, the largest number was sent by the Islamist parties of Pakistan, in particular the *Jamat-i-Islami* (Islamic party). Youth affiliated with *Ikhawan Ul-Muslimeen* (Muslim Brotherhood) from the Middle Eastern countries and Far East came to wage holy war. The total number of all these groups was never more than a few thousands at any time. Interviews with Afghan party leaders in Peshawar, January 1990.

46 The Kabul regime made its first attempt to remove the ideological dimension from its conflict with the Mujahideen in December 1986. This included dropping the word 'democratic' from the title of the Republic of Afghanistan, backtracking on its communist identity, and reconciliation with the opposition. See details in Republic of Afghanistan, *National Reconciliation: Documents,* (Kabul: State Printing House, January, 1987).

47 Rasul Bakhsh Rais, 'Afghanistan After the Soviet Withdrawal', *Current History,* Vol. 91, No. 563, March, 1992, pp. 117-22.

48 The relationship between the US and Russia has yet to define itself, but rivalry, confrontation and strategic challenges are no longer relevant to their policies today. Washington shares with the current leadership of Russia an interest in rebuilding the Russian economy and, by doing so, in strengthening the democratic forces there. Since the Gulf war Moscow has played a supportive role, and the US does not see Moscow as a threat. See a report by Patrick E Tyler, 'U.S. Strategy Plan Calls for Insuring No Rivals Develop,' *New York Times,* 8 March 1992, pp. A1, 14.

49 Barbara Crossette, 'U.S.- Moscow Pact on Kabul Clears Way for Transitional Rule,' *New York Times,* 2 January 1992.

50 *New York Times,* 26 April 1992, p. 1.

51 It was the failure of the coup attempted by the communist hard-liners in Moscow in August 1991 that began to expose the vulnerability of the Marxist regime of Najibullah because of its almost total dependence on its Soviet patrons. As support from Moscow became uncertain with the political ascendancy of Boris Yeltsin and the break up of the Soviet Union, the Uzbek and Tajik military commanders stationed in the north became more sceptical about the survivability

of Najibullah. Soon they formed secret alliance with the Mujahideen commanders, in particular Ahmad Shah Masud, which precipitated the downfall of Najibullah. See a report by Barbara Crossette, 'Defections Deepening Kabul's Isolation,' *New York Times,* 9 April 1992, p. 3.

52 Foreign Assistance Act of 1961, Section 620E (e).

53 Edward A Gargan, 'Fiscal and Political Forces Move Pakistan to Seek Afghan Peace,' *New York Times,* 16 February 1992, p. A1.

Chapter 2

Revolution, Reforms, and Rebellion

The Saur revolution of 1978 introduced a new era of Marxist-oriented radical reforms in Afghanistan that led to disruptive changes in the social, economic and political structures of a predominantly feudo-tribal society. The reforms were intended to bring about much needed economic change by abolishing the oligarchical power arrangements that, in the judgment of the Afghan revolutionaries, had prevented progress, development and modernization in the past.[1]

The ideology of the People's Democratic Party of Afghanistan, its quest for political legitimacy, and the desire to win mass support were the factors that necessitated the launching of the radical reforms. But, as the ideological orientation of the new Afghan rulers, the direction of their reforms and their implementation strategies began to unfold, all segments of Afghan society affected by the reforms began to mobilize mass resistance. Characterization of the reforms as alien in inspiration and against the spirit of established Afghan traditions made the calls for resistance widely popular. But a more important reason for the outbreak of the spontaneous revolt was the intrusion of the state into the affairs of an autonomous, though fragmented, Afghan society. This raises the fundamental question of the relationship between the state and society in Afghanistan. The evolution of the Afghan state suggests that its authority over civil society did not extend beyond the legal claims of sovereignty over the subjects. Whenever a tribe or a section of the population revolted against the rulers, the state briefly intervened through the armed forces and quickly withdrew after obtaining fresh allegiance. Quite often conflicts with the tribal society resulted from the efforts of the state to extend its control over a wider section of the population.

The state and tribal society existed as two rival centres of political authority. Harmony between them could be assured only by allowing the rural communities to be governed by local institutions, which were dominated by the land-owning class, tribal chiefs, *khans* and *maliks*. Conversely, the state had only limited resources to compete for influence in the rural areas and to penetrate society at grass-roots level, other than by maintaining patron-client relations with the local elites. Lacking both material resources and political capacity, the Afghan state had long restricted itself to administering Kabul and the provincial centres. By default, the important functions of state were left in the hands of the rural elites.

Each tribe and ethnic community identified itself more with the local political order than with the central state. The localized traditional institutions exercised complete authority over the individual in their respective jurisdictions. This order was sanctified by tribal custom and tradition, and was perhaps the only political entity that the majority of the rural Afghans had known and respected. Although inequalities in the social and economic status of the rural elites and the peasantry did exist, their relationship was free of class-consciousness or a class conflict. The tribal social order is not strictly hierarchical or impersonal like that of the modern state. Therefore, the social influence emanating from blood ties and interpersonal interactions have been more powerful elements of social control than coercion or the threat of force. The rural peasants never saw the system of local authority as exploitative or imposed upon them against their will.

The absence of the state in the countryside strengthened the traditional bonds of dependency of the peasantry on the *khans, maliks* and tribal chiefs. Apart from the traditional ties, ownership of the economic resources, such as land and water, was an important factor in further strengthening the power of the local elites. They more or less substituted for the agencies of the state in providing employment on their lands and physical security against rival tribes. The landowning elites symbolized political power and there was hardly any challenge to their authority.

The *Khalqis* captured a state which had existed on the fringes of Afghan society and when they attempted to penetrate it through reforms, they found they had to compete with the old rivalry of powerful sections of the rural elites. The peasantry, which was the intended beneficiary of the systemic change and redistributive policies, were never on the side of the state in the ensuing power struggle. The *khalqis* carried an overload of their offensive ideology and links with the Soviet Union, which made it doubly difficult to win the support of the rural communities, who rebelled and swelled the ranks of the resistance against the Marxist state.

No less provocative was the conduct of the PDPA in power. After having assumed control of the state apparatus, the party leaders used the well-beaten rhetoric of 'enemies of the revolution' to suppress opposition. Using detentions, torture, depriving the dissidents of their property and plain murders of political rivals,[2] the party fomented the rebellion. Whatever political consensus and institutions had held the Afghan polity together began to disintegrate.

The destabilizing effects of the Saur 'revolution' that provoked a national uprising and the consequent Soviet intervention have raised many questions about the character of the revolution and its leaders. Was the Saur revolution instigated by the Soviet Union? Did the Soviet Union play any direct or indirect role? What were the political and social bases of the revolution? There are three aspects of the revolution that have made the scientific analysis of these and other questions immensely complex. First, the leftist movement in Afghanistan was rudimentary in organization and had virtually no popular support.[3] Its ideological appeal was negligible. Second, the revolution was initiated by a group of Soviet-trained military officers who secretly collaborated with the PDPA leaders.[4] It was more of a military *coup* than a genuine revolution resulting from a 'proletarian struggle' or from a mass movement. Third, the widely-suspected Soviet involvement in the launching and success of the *coup* undermined the legitimacy of the leftist government.

Before we begin to examine various aspects of the origin and character of the Saur 'revolution', it is necessary that we state our premises from the outset. First, the Saur revolution was primarily an internal political development in which only local actors participated although the Soviets undoubtedly knew the plans of the Saur revolutionaries and watched developments carefully. The Soviets naturally did not lack interest and probably encouraged the Khalq and Parcham leaders to stage the *coup*.[5] The Soviet Union had grown disillusioned and weary of Daud's tilt toward Iran and Saudi Arabia after 1975. It was widely perceived that Daud was attempting to balance the Soviet influence in Afghanistan with the help of pro-Western Islamic countries, and through them possibly the US.[6] The shift in foreign policy was one of the factors that had earlier estranged *Parchamis* from him.

Secondly, whatever the *modus operandi* of the revolution and the role of the military, it was Marxist-Leninist in inspiration. In the first few months of the political change, the new regime claimed itself to be nationalist and committed to defending the principles of Islam, democracy and freedom.[7] Lacking mass support, it was necessary for the party to identify itself with popular national sentiments. It attempted to cultivate an impression of having formed a national government representing all sections of society ranging from peasants to capitalists and the patriotic *ulema* (scholars of Islam).[8] It was not long before the PDPA leaders began to compete with each other in revolutionary phrase-mongering. Increasing rhetoric on the 'epoch-making ideology of the working class'[9] revealed the true character of the party. In the beginning, attempts of the PDPA to make socialism palatable to the common man under the guise of reforms were sensible but the Islamist opposition to the party knew its origin and foreign connections too well and lost no time in exposing its true ideological orientation.

THE LEFTIST MOVEMENT IN AFGHANISTAN

Although the origin and activity of leftist politics in Afghanistan may be traced back to the organization of *Wikh-e-Zalmayaṇ* (Awakened Youth) in 1947,[10] the parliamentary elections of 1949 for the first time brought to the fore a large group of leftist-leaning parliamentarians. More organized and ideologically motivated criticism of the Afghan power structure began with the introduction of the 1964 constitution.[11] After the end of Daud's autocratic decade, 1953-1963, during which he established an iron rule, open Afghan politics witnessed a refreshing start when political activities were allowed. Although the Afghan monarchy retained all vital powers of state, the 1964 constitution made provision for elections to the parliament. That marked the beginning of a relatively free debate on national issues. Liberal press laws gave the political groups freedom to publish newspapers and magazines to articulate their points of view.[12] As a result of electoral politics, parliamentary procedures and a free Press, a wide array of informal political organizations began to emerge. The ideological orientations and national programmes of these groups ranged widely from Islamism to Marxism-Leninism.

However, the traditional Afghan power structure remained unaffected by both the free debates in the parliament and its criticism in the Press. The oligarchical power structure which centered around the monarchy remained stable. Out of the 216 elected members, 146 were tribal leaders. The *ulema* constituted the largest professional group.[13] The feudal, bureaucratic, religious and tribal elites were integrated into the system through traditional linkages and patron-client relationships. They supported the monarchy not only because it was a symbol of Afghan unity but also because it adequately served their interests.

Taking advantage of the relatively liberal political phenomena, the Afghan Marxists who had worked either underground or individually among students and labour unions, mostly in the urban areas, decided to launch the PDPA in 1965.[14] Nur Muhammad Taraki was unanimously elected by the founders of the party as General Secretary and Babrak Karmal as the First Secretary. Mir Khyber Khan and Hafizullah Amin, though not present at the founding congress, were among the top leaders of a tiny PDPA in the early years. In fact, PDPA was more or less the merger of several socialist groups that had different ideological leanings. Personal differences between their leaders were temporarily subdued by what appeared to be a common desire to unify the revolutionary movement.

After a short period of only one and half years of unity, the PDPA split along two major group affiliations, Khalq and Parcham. The falling apart of the PDPA alliance in 1966 had less to do with profound political differences than with the political divisions which are common in Third World societies and are characteristic of political fragmentation in Afghanistan. Between the two groups, ethnic and leadership factors played divisive roles. Personality clashes and claims to leadership of the leftist movement also widened the gap between Khalq and Parcham. The predominantly Pushtun identity of Khalq and the rural social background of its leaders was in contrast to the well-connected and relatively urban and non-Pushtun characteristics of the Parcham leadership.[15] Although both of these groups considered the Soviet Union their mentor and the true model of socialist reconstruction, the Parcham leaders favoured the involvement of and collaboration with the northern neighbour in order to capture power and implement revolutionary changes while the Khalq leaders, though not averse to co-operation with the Soviet Union, advocated an Afghan pattern of socialism that would not necessarily be offensive to the Soviet Union.

It is equally important to note that their doctrinal differences on revolutionary strategy were pronounced. Khalq leaders preferred a mass organization approach with specific emphasis on class line and working people. The *Parchamis* considered that approach unrealistic in view of the absence of a large industrial proletariat class with an illiterate and extremely conservative peasantry. Karmal preferred a united front of all the progressive Afghan elements including the national bourgeoisie as being more pragmatic and sensible than the dogmatic class line of the *Khalqis*.[16]

After the break up of the PDPA, two additional socialist groups under the names of Shula-e-Javed (eternal flame) and Sitam-e-Milli (Against National Oppression) emerged, led respectively by Dr Rahim Mahmudi and Tahir Badakhshi. While Shula-e-Javed split away from Parcham, Sitam-e-Milli was a new group that for a long time had conducted its own study circles. Both of these groups had pro-Peking orientations and were small in numbers as compared to either Khalq or Parcham. The Sitam-e-Milli favoured a Maoist approach to revolution and began to focus on organizing a peasants' movement.[17] There is no evidence that this group was ever able to do any solid political work. Although Shula-e-Javed and Sitam-e-Milli survived the post-revolutionary oppression, they were never able to offer a viable alternative to the ideology and programme of the PDPA.

Over the years their cadres have joined either one of the several resistance groups or the Kabul regime.

It is interesting to note that all the socialist groups have called themselves after their official newspapers. There are two reasons for this. First, while the press law which liberalized publication of newspapers was enforced in 1965, the legislative measure to allow the establishment of political parties was withheld by the monarchy.[18] Secondly, the papers articulated the political views and ideological lines of the groups which published them. This established the groups' ideological identity with the publications, thus making it easy to operate under familiar names after the ban on their newspapers.[19]

In the 1960s, the leftist groups accepted the participatory framework of constitutional democracy, though limited, and focused their political activities in three areas. First, the Marxists actively participated in parliamentary debates demanding radical changes in the economic, political and social structures of Afghanistan.[20] During these debates Babrak Karmal, Dr Anahita Ratebzad and other leftists confronted the conservative and oligarchical elite. Parliament provided the leftists with a public forum to criticize the monarchy and articulate alternative programmes, but due to their small number in Parliament they were unable to influence the legislative process, and discriminatory coverage in the media effectively thwarted their attempts to win popular support for their politics. Second, Parcham and Khalq organized student and labour unions mainly in Kabul and Qanduz: Kabul University's student union was dominated by the leftists till the late sixties. Third, using their support among student and labour unions in urban areas, the leftists organized effective street demonstrations to press their demands for reforms. The agitational politics of the left, though localized to Kabul, did raise concerns of a serious threat to law and order in the capital. Perhaps it was the spectre of disruptive agitational politics of both the left and the right that prevented the monarchy from going ahead with its initial programme of liberal politics. Meantime, though Khalq and Parcham began to recruit members of the armed forces, the bureaucracy and teachers in the countryside, they never attempted to launch a mass organization through either mass contacts or by setting up party units at regional or village levels.

Khalq and Parcham's revolutionary strategy was expedient and well in line with Leninist tradition in using army disaffection in their wresting of power.[21] The reasons for the short cut were compelling. First, a revolutionary change in the classical sense would have had to await a genuine class formation. This was not likely to happen in the lifetime of the PDPA leaders. Secondly, though a glaring division of haves and have-nots on the Third World pattern existed in Afghan society, Khalq and Parcham could neither sharpen the polarization nor immediately exploit it for their political ascendency. This was due to both the urban orientation of the party networks and the power of the tribal elites in the countryside. Nor did the vast peasantry appear inclined to challenge the existing property relations or the traditional structure of authority. The traditional Afghan political culture and social norms so widely respected by rural society presented serious difficulties for the *Khalqis* in their attempt to bring about a change in the Afghan mind-set. The semi-closed nature of the polity was another factor that did not allow them an opportunity to broaden the appeal of their message. Finally, a tribal society inherently resistant to change compounded the problems for a mass-oriented revolutionary movement. Their analysis of Afghan society and self-appraisal had informed the leftists much earlier not only that the rural communities were politically inaccessible but also that it would be a Herculean task to mobilize the Afghan peasantry in support of their ideology. Absence of an urban working class was another factor which restricted Marxist ideology to the study circles of the urban middle-class ideologues and intellectuals. These chair-bound revolutionaries lacked the vision and political capacity to organize the peasants and prepare them for a revolutionary programme. They rather opted for an expedient political course in which the forging of comradeship with the officer corps of the armed forces appeared an attractive alternative.

THE ISLAMIST CHALLENGE

As the agitational politics of the left began to demonstrate its disruptive potential, the traditional religious elite and the Islamists became politically more active and assertive in their demands for Islamization. During the constitutional era, all varieties of political groups attempted to influence government policies, urging reformation of the Afghan society to fit their particular visions. The Islamists and traditional *ulema* exerted pressure on the monarchy to ban the political activities of the leftists which they suspected were being instigated by the Soviet Union. Many of the rightist groups feared, and rightly so, that the increasing Soviet influence not only would encourage the leftists to subvert the traditional political process but also carried the risk of covert Soviet support to them.

The debates in the parliament and a brief period of uncensored publications showed up a sharpening of the left-right polarization in Afghan politics. This ideological divide, which is pervasive in Third World politics, generated mutual fears and antagonism. Not accustomed to the values of a pluralistic political culture, the groups on the opposite sides of the ideological rift sought to eliminate the others by whatever means available.

The Afghan right was alerted to the activities of the left and responded in many ways. They counselled the King to ban the leftist publications such as *Khalq* and *Parcham,* and they succeeded in achieving this objective. The oligarchical elite that had traditionally clustered around the monarchy shared the perceptions of the conservative *ulema* and supported their view of the leftist threat. Unlike the Islamists, the traditional *ulema* did not turn hostile to the King or his patron-client system.[22] Most of them argued instead that the monarchy was Islamic in its character and that the institution could be reformed and entrusted with the further Islamization of Afghan society.

To counter the activities of the leftist groups, the Islamists began to set up their political fronts among students and factory workers. The nucleus of Islamists, largely influenced by the Ikhwan-Ul-Muslimun (Muslim Brotherhood) of Egypt, the Jamat-e-Islami of Pakistan and the Islamic movement in Iran, organized Hizb-e-Islami in Kabul. Their concentration and focus on Kabul University students brought results. Within a short period of time, the Islamist students forced out the leftist student groups. In 1972, the Islamists won two-thirds of the elected seats in the Kabul University student union.

This encouraged and motivated the Islamists to organize themselves on a national basis by recruiting like-minded *ulema* and students of Islamic theology from the provinces. However, before they could emerge as a formidable political force, the military *coup* which brought Daud to power in 1973 severely limited their activities. Both the Islamists and conservative *ulema* proved that because of the popular appeal of the religious ethos they had greater potential for street demonstrations than the leftists.

The 1969 election, which was the second and the last of the constitutional period, drastically changed the power equilibrium in the Parliament. The rising influence and political activism of the *ulema* was seen both within the Parliament in their greater representation, and outside in their street demonstrations. The representation of the leftist groups was reduced in the House. Only two of the prominent socialist leaders were able to get re-elected.[23] The dominance of the feudal, oligarchical elite and *ulema* in the Parliament further increased conservative influence on the monarchy.

The monarchy's traditional weaknesses of drift, procrastination and indecisiveness were compounded by conservative pressures for immediate action against the communists. *Ulema* demanded concrete steps for the Islamization of Afghan values which in their view had become perverted by modern education. The *ulema* began to demonstrate their strength in the streets of Kabul and other major cities. Their demonstrations targeted communists and the modernist elite occupying high positions in the King's bureaucracy. Both the modernization programme, though limited, and leftist activities were actively opposed by the *ulema* as a threat to Islam and to Afghan tradition.

THE 1973 COUP

In July 1973, while King Zahir Shah was relaxing on the Italian island of Ischia, his cousin and brother-in-law, Sardar Mohammad Daud, proclaimed the end of the monarchy. He designated himself President of the Republic of Afghanistan and announced that he had led the *coup* to replace the 'corrupt and effete' rule of Zahir Shah with 'genuine democracy.'

The 1973 *coup* raised many important questions. Why was a *coup* staged on the eve of the third parliamentary general elections under the 1964 constitution? What was the nature of the alliance of different political and bureaucratic groups that successfully engineered the *coup?* Was it a concerted response to the rising power of the Islamists or the result of Sardar Daud's personal power ambitions? Why did the leftists co-operate with Daud or why did Daud co-opt them? What were the motivations and objectives of the different forces involved in this *coup* which proved a decisive turning point in Afghanistan's political history?

The explanations of the 1973 *coup* are diverse and contradictory. One of the widely held views was that Sardar Daud had felt deprived of an opportunity to share power. The 1964 constitution had barred all members of the royal family from participating in politics. During his premiership during the decade 1953-63, Daud, though an autocratic ruler, had earned tremendous recognition as a nationalist and a capable modernist. One reason for his tilt towards Moscow was the desire for rapid development and modernization of the Afghan economy and armed forces.[25] Not surprisingly, Daud found much support among a large section of the military which he had modernized and equipped with Soviet assistance during his tenure.[26] Daud used his influence and image as a modern ruler to stage a comeback through the support of military officers who had grown impatient due to Zahir Shah's conservative policies. An analysis of contending political ideologies, group politics and increasing pressures on the monarchy from all sides for reforms would suggest that the motivations for the *coup* as well as the objectives of its leaders were more diverse and complex than Daud's personal ambitions for ascendance to power. Rather, a broad coalition of political and bureaucratic forces collaborated with him for almost identical reasons to end the monarchy.

There were three elements in this coalition that threw its support behind Daud. They may roughly be characterized as nationalists, modernists and socialists. This identification is more due to the group orientations of these elements than to strict compartmentalization of their categories. Their semantic division into socialists, modernists and nationalists becomes misleading when one considers the practical aspects—a socialist can be both a nationalist and a modernist. These categories have been chosen only to indicate the traditional constituencies of various groups who were commonly labelled as such. Sardar Daud was acceptable to the Afghan Army, the civil bureaucracy, secular intellectuals of the middle class and the Marxists, particularly, Babrak Karmal's Parcham, not only because he was popular and had strong credentials as a nationalist but also because he had a charismatic personal appeal.

Among the many causative factors that produced the 1973 change, two might be mentioned as crucial. First, the partial implementation of the 1964 constitution had not changed the absolute character of the monarchy. Although the King allowed elections and free parliamentary debates, he did not budge on conceding any real power to the elected representatives of the people. He withheld constitutional provisions for the establishment of political parties, frequently vetoed legislation and obstructed any meaningful change in the power structure. The electoral process neither helped bring into the Parliament the change-oriented modernists nor did it achieve any success with the detribilization of the political process. It simply institutionalized the representative status of the members of the Afghan oligarchy. Moreover, the composition of the parliament held out no promise of becoming a democratic institution, because individual members had no propensity or desire for reforms. Domination of the parliament by the traditional elite frustrated the members of the newly emergent educated middle class. The gradual democratization of Afghan society envisaged in the promulgation of the 1964 constitution rather seemed to be strengthening the hold of the oligarchy. This increased the disillusionment of all varieties of modernists concerning the prospects for change and social mobility in Afghan society.

The second factor relates to the growing influence of both the conservative *ulema* and the Islamists. The modernist sections of the urban elite, the army, the civil bureaucracy and secular intellectuals perceived a potential danger in the rising power of the Islamists in particular. The organizational pattern of the Islamists and their political activism, which they demonstrated in the streets of Kabul, posed a threat to the gains that the urban coalition had made in terms of setting the state on the course of modernization, particularly in the fifties and sixties. More immediately, the growing political strength of the Islamist forces opened the monarchy up to the pressure of the conservative elements. In order to accommodate the conservative elements the King began to deviate from the promised reforms and the pursuit of liberal policies after the second parliamentary elections in 1969. Also, he came to share the view of the oligarchical elite around him that further democratization of the system would destabilize the status quo power arrangements.

The coalition that supported Daud's seizure of power was shaped by consensus on four central issues. First was the common vision of a modern Afghanistan. Second was the common perception of a threat in the rise of Islamist elements. Third, an obstructionist Parliament, dominated by conservative elements, showed no promise of reforms, development or any change in the political hierarchy. Finally, the perpetuation of the monarchy was seen as instrumental to the stability of the oligarchical political order. This seemed to be blocking the entry of the new middle class and professional groups into important roles in the government.

Daud and his allies decided to capture power through military intervention in order to end the monarchy and shape Afghanistan's politics in their own vision. Military action was engineered by a small group of military officers who were secret members of Parcham.[27] A takeover plan was implemented by a few military officers, among whom Aslam Watanyar and Abdul Qadir played the central role. There was no resistance worth the name; it was almost a bloodless *putsch*.[28]

After coming into power Daud announced several fundamental changes in the old political order. He declared Afghanistan a republic, abrogated the 1964 constitution, dissolved the parliament and promised reforms and the promulgation of a truly democratic constitution.[29]

Parcham's involvement in the *coup* and support for Daud was expedient. First, Karmal and other *Parchamis* thought that by helping to end the monarchy and demolishing its oligarchical power structure, they could bring Afghanistan to a 'national democratic revolutionary' stage. Khalq took a polemical position, characterizing the event as a palace *coup* and Parcham's power-sharing with Daud as a betrayal of the true revolutionary line. Second, Karmal assessed that as Daud had no party of his own, Parcham could emerge as a legitimate force within the power establishment of Afghanistan. The scheme to cultivate Daud's dependence on Parcham, along with his old age and failing health, strengthened Karmal's hopes for political ascendency. Third, the *Parchamis* calculated that their integration into the power structure would make available to them the resources to organize the party better and to attract more cadres from other socialist groups. Finally, mending fences with its traditional political rivals was another factor that motivated Parcham to co-operate with Daud.

In the initial phase, Parcham formed the inner circle around Daud. He appointed some of its members to high administrative and ministerial positions. It was no accident that Parcham acquired control of the Ministry of the Interior, which supervised the police and intelligence networks in Afghanistan.[30] In the first move, Parcham recruited about two hundred of its activists as police officers.[31] Its control of the police forces facilitated its hunt for the Islamists who had opposed the *coup* and Parcham's involvement as a prelude to greater Soviet influence in Afghanistan. The rank and file of the conspirators from the Afghan Army who were instrumental in pulling off the *coup* were promoted to higher positions in the government. This also strengthened the clout of Parcham. The *Parchamis* began to assert themselves in the appointments and transfers of high-ranking officers in the Afghan bureaucracy.[32] The increasing influence of Parcham in government affairs eventually frightened Daud as he felt isolated from both the army and the old oligarchy.

DAUD'S PROBLEMS

Sardar Daud's decision to collaborate with Parcham and to assume leadership of the coalition which was brought into power by the armed forces reflected certain misperceptions on his part about the power-base of his partners and his own ability to control them. While deeming collaboration with Parcham and officers of the

Afghan Army necessary for capturing power, Daud might have thought that he would eventually get rid of them. Therefore, Daud's shift from his traditional constituency of feudal and tribal oligarchs to the Marxists and men in uniform was motivated more by expediency than by any ideological affinity with them. Daud assumed that, once in power, he would effectively use political patronage to muster enough political support among the members of the Afghan oligarchy and educated elite to subvert the power of his leftist allies.

Two incidents opened his eyes and made him realize, perhaps for the first time, how fragile was his power-base. First was the mysterious death of Mohammad Hashim Maiwandwal, a former prime minister, in his prison cell in October 1973. He was implicated in a *coup* attempt and was awaiting his trial. Contrary to the official report that he 'committed suicide', he was widely believed to have been killed by the leftists. They did this, perhaps, without the prior consent of Daud. The reason why the leftists may have wanted to eliminate Maiwandwal was that he could be a formidable political rival, given his political experience and international standing.[33] Also, Maiwandwal could be one of Daud's potential allies against the Marxists. Being aware of Soviet influence and interests in the leftist movement in Afghanistan, Sardar Daud moved very cautiously against Parcham.[34] He quietly started removing the hard-line leftists from their governmental positions and shuffling around those whom he thought would not pose any threat.

The second incident was the uncovering of a conspiracy allegedly hatched by the leftist army officers in September 1975 to overthrow Daud's government. About forty officers were arrested for plotting against the republic. The fact that Daud did not execute any of them, or even pursue the matter more seriously than simply dismissing the officers, suggests that he might have contrived or used the event to purge *Parchami* members in the armed forces. After this he turned to the Parcham appointees in the civil administration and removed a large number of them. He was careful not to ascribe ideological reasons for dismissing the *Parchamis*. Although Parcham, and probably the Soviets, were aware of the real reasons for the removal of *Parchami* officers, Daud always cited corruption and inefficiency as reasons for their transfers and dismissals.[35]

After breaking away from Parcham, Daud began reviving links with the moderate and even Islamist groups for reconciliation and political support. His shift towards the Afghan establishment was, however, too late and lacked any credibility. The Islamists had been antagonized by both the co-option of leftists and persecution by them. The oligarchy did not trust him either due to his dissolution of the monarchy. Though he withheld the promised land reforms to catch the attention of the feudal and tribal chiefs, he could not regain their support, which he had lost considerably by aligning himself with Parcham. The educated Afghans were frustrated by his autocratic policies. The promised 'true democracy' was never introduced. The enforcement of the 1977 constitution rather ended any hopes of freedom, democracy, the revival of political parties and popular representation.

The structure of the republican state which the constitution outlined concentrated all powers in the office of the president. It was a formalization of the traditional authority of the Afghan monarch in the guise of parliament, elections and representative government. The Milli Jirga (National Council) became a single-house legislature, replacing the former bicameral parliament. The National Council was so powerless that it could not even introduce legal bills, and could merely act on the bills proposed by the government. The constitution allowed only the official party, the Hizb-e-Inqilab-e-Milli (National Revolutionary Party) to function. This party headed by Daud controlled the National Council. Both in theory and practice, the executive was not accountable to the Parliament but to the president.

The president was the repository of all the powers of the state. He was a chief executive without any check on his authority to run the government. The constitution empowered him to dissolve the Parliament, call new elections, appoint judges, declare an emergency, suspend civil rights and act as the commander-in-chief of the armed forces. The centralization of state power and its concentration in the office of the president turned the Afghan political order into a one-man show. The objective was not only to enhance the personal powers of Daud but also to restructure the Afghan state in a fashion that would allow Kabul a greater capacity to expand and modernize.

The strengthening of the state had to precede the state-centred modernization. However, in practice, each complemented and reinforced the other. Measures such as centralized planning, nationalization of the financial institutions, government control over exports and industrialization in the public sector were meant to generate resources for expanding the role of the state in a traditional society.[36] The agricultural economy also received a tremendous push from the government. The construction of roads and irrigation networks[37], largely through foreign assistance, had already paved the way to inter-regional trade, and in some areas commercial agriculture had begun to develop. Daud's republic launched some very ambitious agricultural projects. It established government owned farms, which were thought to have demonstrative effects and also increased production. Extension services and low-interest loans by the government's Agricultural Development Bank were aimed at encouraging the mechanization of agriculture. Private investment in the commercialization of agriculture also increased with high returns on the export of agricultural products. Money-lenders, who expected regular repayment of loans with higher interest yields, appeared more forthcoming in giving loans.[38]

The development paradigm based on the centralization of political power and state-supervised economic development did not fit well into the social conditions of Afghanistan. It compounded the old dilemma of balancing power between the central government and the regional elites. While benefitting from the government-supported agricultural development programmes, they remained resistant to any encroachment on their authority. In the long run, however, changes in the rural economy might have caused a rearrangement of power relations between the central government and the rural elites on the one hand, and the traditional elites and the urban-based commercial agriculturists on the other.[39]

The commercialization of agriculture, although still fragmented and restricted to small areas, produced new economic dislocations and inequalities of income. Rich farmers and large landholders grew richer with the use of modern agricultural implements, while the peasants began to proletarianize. In some areas, heavy debts incurred in the past, and impoverishment caused by the modernization of agriculture, resulted in the loss of small pieces of land owned by the peasants to land-hungry money-lenders. In certain areas the creditors increasingly used loans as a bait to small landowners 'to gradually acquire the land itself'.[40] But it would be wrong to conclude that agriculture was fully commercialized. The subsistence sector still accounted for 65 to 70 per cent of agricultural production.[41] By the middle of the 1970s, modern farming methods and market orientation had only begun to reach the rural areas. Agricultural development did not, on the whole, achieve a very high level. Marginal increases in productivity did not bring prosperity to the agricultural sector. Burgeoning foreign loans that had financed some of the agricultural projects added to the difficulties of financial management. Moreover, the programmes which Daud had initiated had a long gestation period and it would have been very difficult for him to make any political capital before the results were visible.

Threats to the Daud regime emanated both from the economic difficulties and from uncertain political conditions. Within a period of only a few years he confronted serious political troubles. The increasing belligerency of the conservative political groups and intrigues of the leftists began to undermine his regime. To control opposition to his rule, Daud relied more on repressive measures than on initiating a genuine political reconciliation, dialogue or accommodation with the opposition political groups. One of Daud's failings was that he never realized the fact that the constituency that had supported him during his premiership had either become disillusioned with him or had little motivation to mobilize political support for him. The use of extensive persecution, arbitrary imprisonments and even assassinations of political opponents, both of the right and the left[42], proved highly ineffective in controlling or even minimizing political dissent.

In the wake of formidable opposition at home and Soviet suspicions of him, Daud began to reorient Afghanistan's foreign policy by emphasizing 'true non-alignment,' 'genuine independence,' and *rapprochement* with the neighbours.[43] The shift in foreign policy, though moderate and primarily in line with Soviet sensitivities, had some important domestic political objectives. First, Daud and his advisors thought that large-scale foreign aid, particularly from

Iran and Saudi Arabia, not only was required to save the rapidly deteriorating economic conditions but also could produce favourable political effects in winning the sympathies of the traditionalist elite. Agricultural and industrial projects were expected to be funded by the Islamic countries which, in the assessment of his advisors, could generate employment and economic activity, thus garnering political support for the policies of the regime. Secondly, Daud perhaps needed external co-operation both to avert the leftist threat as well as to balance the Soviet influence in Afghanistan.

Daud's domestic political compulsions coincided with the Shah of Iran's ambitions in the mid-seventies to play a larger security role in the region under the US umbrella. The Shah's desire to contest the rising Soviet influence and his willingness to support regimes attempting to suppress pro-Moscow sub-nationalist and socialist elements in the region attracted Daud to an Iranian-centred regional balance of power more for domestic than other motivations.

The Shah supported the Bhutto regime in Pakistan by various means, both material and political, to suppress the Baloch rebellion of 1973-7. Iranian troops were actively participating in the suppression of the Douhfery rebellion in Oman. An increasing arsenal supported long-term plans to make Iran a formidable power, and encouraged the Shah to undertake such adventures.[44]

Daud's shift toward the regional Islamic countries was reflected in a *rapprochement* with Pakistan brought about by the diplomatic efforts of Iran.[45] Daud not only compromised Afghanistan's traditional support for Pushtunistan and changed its position on the Durand Line but he also secretly agreed to eject the Baloch guerrillas who had been operating from Afghan territory.[46] To compensate for the likely fall in Soviet aid as a result of the reorientation of Daud's foreign policy, and to further induce him into a Tehran-centred regional security set-up, Iran promised to provide $2 billion in economic assistance, accompanied by a commitment to grant transit facilities through Iran's territory. Likewise, Saudi Arabia made a pledge of $500 million in economic assistance.[47]

While relations with the Soviet Union remained normal and Daud avoided hostile posturing, the easing out of Parcham from the government and the increasing Iranian influence raised Soviet doubts.

Moscow was no longer certain about Daud's foreign policy moves. This was partly due to Daud's failure to communicate to the Soviets that he never contemplated abandoning Afghanistan's traditional neutrality and non-alignment. When Daud visited Moscow in 1977, he was reportedly told of Soviet displeasure over his obvious tilt toward the Islamic states of the region.[48] His encounter with Brezhnev during this last visit to Moscow convinced him that the Soviet leader was planning to make trouble for him through the Marxist groups. Fearing covert Soviet support to the Marxists, Daud began to lean ever more toward Iran for intelligence support, which in turn further strengthened the Soviet fears and distrust.[49]

UNIFICATION OF THE PDPA

Daud's domestic and foreign policies greatly influenced the unification process of the PDPA. Although it is often argued that the Soviet Union played a central role in bringing about a *rapprochement* between Khalq and Parcham in 1977, the domestic political scene is conveniently ignored. Given their wide contacts the Soviets might have exercised their influence over the Afghan Marxists to forge unity. The communist parties in the region, notably the Communist Party of India, counselled Parcham and Khalq to sink their differences and close their ranks. This did help the initiation of a dialogue between them[50] but in the final analysis, it was their common fear that Daud, with the support of Iran and by turning to the oligarchical elite, was preparing for a final move to destroy the leftist movement.

The uniting of the traditionally feuding Marxist factions reactivated the party among both the armed forces and disaffected sections of the educated middle class. At this point, the party put more emphasis on contact with members of the armed forces than on mass education or political organization. The Afghan Marxists were realistic in this respect. Given the backward nature of the Afghan peasantry and the absence of a large industrial working class, the support of the military was the only means of capturing power. During this period, the Soviet Union appeared to be interested in affirming support to the 'revolutionary democratic states' controlled by the military.[51] The

Marxists, lacking extensive political organization and in the absence of the relevant social formations for a revolution, relied upon the army as the chief instrument of revolutionary change. In the judgment of a prominent Soviet specialist on Afghanistan, R Akhramovitch, this could pave the way for the revolution proper.[52]

THE SAUR REVOLUTION

In many ways, Afghanistan did not appear to be ripe for a Marxist revolution in April 1978.[53] First, the organizational apparatus of the PDPA was weak, divisions along ethnic and personal lines were deep, and its image and ideological line lacked popular appeal in a very Islamic and traditional political environment.

Second, the Soviet Union, the other actor in post-revolution politics, appeared to be cautious and unsure about the strength and effectiveness of the Marxist groups. In its strategy of political penetration, Moscow preferred the cultivation of dependency links with the Afghan ruling elite to outright support of a Marxist take-over of power in Kabul. This proved an effective course in obtaining a privileged position in Afghanistan. The Afghan monarchy, and later the republican regime of Daud, had carefully accommodated Moscow's interests. Co-option of the *Parchamis* by Daud further enhanced Soviet influence. Up until the mid-seventies, the Soviet leaders were satisfied with Kabul's domestic political arrangements and its Moscow-tilted, non-aligned foreign policy. Therefore, the Soviets carefully avoided any extensive and visible support to the communist movement. It was only after they feared that Daud was introducing external influences in Afghanistan which were hostile to the Soviet Union and could possibly result in its 'expulsion' from the country that the Soviets began to give serious thought to the Khalq-Parcham alternative to counter Daud and, if possible, remove him from power.

Daud's break with Parcham coupled with his new foreign policy initiatives heightened Soviet concerns. His overtures toward the Islamic countries in the region to obtain economic aid and transit facilities may not have raised Soviet ire had the Soviets not read something suspicious in his intentions. They understood well Afghanistan's need for external aid, which they could not increase to a higher level than they were already providing. However, Moscow was perturbed both by the increasing influence of the Iranian intelligence agencies and by Daud's statements criticizing Soviet allies Cuba and Ethiopia for deviating from true non-alignment.[54] In yet another major policy shift, Daud decided to send Afghan military officers to India and Egypt for training. Daud perhaps did not take the Soviets into his confidence in mending fences with Pakistan either.[55] More than anything, the Soviets were overly sensitive to and apprehensive of the regional ambitions of the Shah of Iran, who they thought was attempting to destabilize their privileged position in Afghanistan.[56]

On the surface it would appear that the reorientation of Daud's foreign policy and changes in his government's position on domestic issues did not ruffle the Soviet leaders enough to warrant his overthrow. Return visits and joint communiques did not reveal much about the cooling off of relations between Kabul and Moscow, but a closer look at the domestic political scene in Afghanistan would suggest that Moscow was growing wary of Daud's policies. There was enough ground for their suspicion. First, he eased out the leftist groups from the government and appeared determined to crush them. This altered Daud's image as a friend of the 'progressive forces', which the co-opting of Parcham had earlier established. Second, he began removing the Soviet advisers who were attached to the Afghan army units. Third, he seemed to be asserting more independence from Moscow and hardly caring for Soviet 'advice', which he regarded as an interference in the internal affairs of his country.[57] How much of a role the Soviet Union played in organizing the *coup* still remains a matter of controversy, but circumstantial evidence abounds pointing to its interest in removing Daud. First there was the much-publicized displeasure of Brezhnev with Daud over his domestic and foreign policy initiatives. Second, there were Moscow's covert links with the Afghan Marxists who overthrew Daud. Third, there was the immediate assistance and co-operation which Moscow extended to stabilize the Marxist regime, which later further deepened her involvement and led to the military intervention.

The Saur revolution of Afghanistan was unique in one respect. The military *coup,* unlike the praetorianism in the Third World, was guided and supervised by the civilian leaders of the PDPA.[58] This explains the PDPA's penetration of the Afghan armed forces and its 'vanguard' position. The Parcham and Khalq leaders had developed ideological links with the Soviet-trained Afghan army officers since the late sixties. Besides, ideological romanticism and a common desire for change in an oppressive oligarchical power structure facilitated comradeship among civilian Marxists and their collaborators in the armed forces. The majority of them were inspired by the Soviet model of revolutionary change and socialist reconstruction. Although pro-Beijing and Maoist groups did exist in Afghanistan, the overwhelming majority of the Afghan leftists favoured a pro-Moscow line for political and strategic reasons.

With hindsight, the planning of the Saur *coup* would suggest that although the PDPA was preparing itself to replace Daud and waiting for an opportune moment to strike, certain domestic events advanced the schedule. Originally, the revolution had been planned for July-August.[59] The murder of Mir Ali Akbar Khyber, an ideologue and veteran Marxist, on 17 April triggered a public outcry which was skilfully organized by the PDPA. The Afghan leftists blamed Daud and SAVAK (the Iranian intelligence agency under the Shah) for Khyber's assassination. The funeral procession attracted a large crowd which showed, in part, the growing resentment against the Daud government. The PDPA leaders used the occasion for making hard-hitting speeches against Daud, vowing to avenge the blood of the 'martyr'. The tone of the speeches and threats, and the wide participation of Kabulis in the funeral procession, which smacked of a political gathering, alarmed Daud. After carefully studying the situation his government cracked down against the Marxists. In the first sweep all top leaders of the PDPA were arrested.

Both the assassination of Mir Ali Akbar Khyber and the ensuing mass arrests convinced the PDPA leaders that procrastination in the *coup* would be suicidal. Several accounts of the Saur revolution plan suggest that Nur Mohammad Taraki's arrest was marked as a signal for staging the *coup.*[60]

Hafizullah Amin was the only top leader who was not arrested the night the others were. Perhaps some *Khalqi* military officers were instrumental in delaying his arrest, and probably revealed to him the plans of the government. Amin fully exploited the opportunity and guided the *coup* from his home before being jailed.[61] His communication with Aslam Watanyar and Abdul Qadir through Syed Mohammad Gulabzoy was crucial in this regard. Those who see a Soviet hand in the *coup* suspect that Gulabzoy, under instructions from the Soviets, not Amin, launched the *coup*.[62] The evidence for this line of argument is fuzzy and highly circumstantial.

As part of their strategy, the PDPA workers staged a protest on 27 April.[63] The protesters marched through the streets of Kabul in relatively large numbers, chanting anti-Daud slogans and demanding the release of detained leaders. About noon Colonel Watanyar led the attack on the presidential palace. The Republican Guards, who were responsible for the security of the President, put up stiff resistance and refused to surrender to the end Qadir led air bombardments over the presidential palace and against the Republican Guards. Kabul radio station and important government and military installations were captured by the rebels in the early phase of the *coup*. Although Daud's defence minister made several attempts to mobilize the loyal troops, he failed dismally in getting them into action. The indifference of the larger part of the Afghan Army officers and the alienation of the troops tilted the balance in favour of the PDPA-led military officers. It is astonishing that the *coup* succeeded with such a limited number of troops participating in the fighting. Daud, his immediate family members and close aides were all shot dead during the takeover of the palace.

The first announcements on Kabul Radio concerning the success of the revolution were made by a military officer, Qadir. He claimed that power had passed into the hands of a military revolutionary council headed by him. However, when the formation of the National Revolutionary Council (NRC) was announced, it was largely dominated by members of the PDPA, drawn from both its Parcham and Khalq factions, and headed by Taraki.[64] The NRC assumed all fundamental powers of the state.[65]

The PDPA regime was quite successful in securing control of the state apparatus in a relatively short period of time. Those officials holding important positions in the Afghan bureaucracy who were suspected of sympathy with the Daud regime were purged. Afghan ambassadors to foreign countries were asked to resign. The administration in the provinces was staffed by loyal army officers. Trusted military men took over as governors. Similarly, changes in the media, educational institutions and other important government organs were swift and sweeping. The removal of old bureaucratic and administrative personnel and their replacement by party workers and army officers was necessary in order to gain effective control of the government machinery. These steps were crucial for a party which lacked mass support. However, the placement of enthusiastic but inexperienced party members into administrative and bureaucratic positions proved disastrous in the implementation of revolutionary reforms and policies, showing the judgmental futility of the whole enterprise.

The early pronouncements of the PDPA leaders revealed very little about the ideological character of their regime. While expressing an urgent need for structural, economic and social reforms, the PDPA leaders put strong emphasis on revolution, nationalism, progress, democracy and respect for Islam.[66] Keeping its ideological line deliberately vague, the regime conveyed the impression that it was essentially reformist.[67] However, a few months later increasingly unrealistic rhetoric on Marxism-Leninism, well-publicized praise of the Soviet system and attacks on Islam and Afghan tradition[68] put the party and its incipient administrative machinery into serious disarray. The power struggle between the Khalq and Parcham factions began to further affect the political capacity of a party that had limited cadres and enjoyed very little support.

Among the various factors that promoted infighting and led to the undoing of the Khalq-Parcham alliance, two might be mentioned as a prelude to a wider conflagration: first, the absence of a sizeable and influential group that could mediate in the conflict within the party, and second, the weaker position of Parcham in the reunified PDPA. The *Parchamis* were half-heartedly accepted, and only as junior partners, by the predominantly Pushtun *Khalqis*, who undoubtedly had a larger political base and greater influence in the armed forces. From the very beginning, Amin and his loyalists in the armed forces and the secret police had resented the appointment of Babrak Karmal, leader of the Parcham faction, as a deputy prime minister.[69] It was a sensible move by Taraki to allocate the second highest position to Karmal alongside Amin. First, by having two deputy prime ministers, Taraki wanted to balance the allocation of power between the Karmal and Amin groups. Second, the power-sharing arrangements with Parcham required that its leaders be placed in respectable positions in the leftist hierarchy.[70]

However Amin and his group, blinded by the traditional Pushtun arrogance, wanted either to allocate to Parcham a subordinate role, or to exclude it entirely from the government. It was largely due to the old antagonism that had developed along ethnic lines and to personality clashes of feudal-minded Marxists that the struggle to monopolize political authority began to dominate their agenda. The difficulties of introducing revolutionary changes also had an impact in exacerbating mistrust and ethnic quarrels within the party. Khalq and Parcham became widely divided sooner than expected on the contents and implementation strategies of the reforms.[71] To some degree, doctrinal differences on pursuing united-front politics, which Karmal favoured and advocated, frustrated attempts to maintain a workable partnership between the two groups.

Amin and his group of comrades, so determined to turn the coalition into Khalq-dominated rule, succeeded in manoeuvering the ouster of Karmal from his number two position in the government. On 5 July 1978, after merely three months of the revolution and after only thirty-five days in office, Karmal and other top leaders of Parcham were sent abroad as ambassadors.[72] An honorable exit, at least by prevailing Afghan standards, was arranged, keeping the door open to a *rapprochement* in future. But, as the strongmen of the Khalq group began to prosecute the remaining *Parchamis* on the grounds of their alleged involvement in a conspiracy to capture power, the PDPA coalition completely fell apart.[73] The inner circle of Amin's supporters in Khalq took over complete control of the state machinery. They reorganized the party, enlarged the central committee and inducted loyalists into the higher party ranks. Amin became Secretary-General of the party and emerged as the most powerful man in the new political set-up, although formally he ranked second to Nur Mohammad Taraki.

With the ouster of Parcham, Khalq became more aggressive in pursuing its radical reforms. However, there is hardly any evidence that Parcham had opposed the reforms. On the contrary, Babrak Karmal later acknowledged that he had participated in enacting them before he was forcibly sent abroad as ambassador.[74] It was the monopoly over political power established by Khalq and the key role played by Amin in government affairs that gave impetus to the implementation of the reforms.

REFORMS

A closer look at the contents of the reforms announced by the Khalq regime between 15 May and 28 November 1978 suggests that it perhaps expected some political support through a rapid process of redistribution and social uplift. The entire question of reforms should be viewed not only in terms of form and content but also in relation to the Afghan tradition and the *modus operandi* of their implementation. While the substance of various reforms was highly desirable and could even be considered fundamental to any process of development in a feudal and tribal society, the political form in which the reforms were articulated contrasted with the popular cultural and social norms of the Afghan masses. The lapses in political communication on the reforms did provide an opportunity to those Afghan groups that were opposed to the regime. Specifically, the landowning and religious establishment feared a loss of economic and social privileges if the reforms were implemented. Before we go any further in discussing the cultural and religious obstacles to the reforms, it would be useful to examine briefly some of those reforms that caused widespread rebellion.

The first set of reforms was aimed at relieving the peasantry of the burden of debt and redistributing land among them by fixing the upper limit on landholding by each family. The Afghan rural agricultural economy had depended largely on money lenders for the financing of seed, equipment and fertilizers. In the absence of any government financing agency with branches in the rural areas, the system worked according to certain norms but tended to be highly exploitative. The private financiers traditionally took the place of agricultural banks in extending loans. The Revolutionary Council approved Decree Number 6 which outlawed this practice and 'waived debts of landless peasants and peasants with small landholdings owed to usurers and mortgagers'.[75] Alternatively, the PDPA regime announced the formation of co-operatives to mobilize funds through the voluntary contributions of the peasants.[76] It was conveniently assumed that funds would be available, seed and fertilizer would be purchased and peasants would be happily contributing to the common agricultural funds. The scheme had serious misconceptions concerning the ability and willingness of the peasants to do things which would otherwise require not only organization but also necessary expertise and government supervision.

The difficulty that Decree Number 6 created was that it put an end to a traditional source of financing without replacing it with a modern economic structure that could enable the peasants to meet their timely requirement of growing crops. The money lenders, having lost money, were not willing to risk more despite assurances from the peasants to pay them back.

Another decree, Number 8, which was issued on 1 December 1978, attempted to eliminate 'feudal and pre-feudal relations from the socio-economic system of the country' by making drastic changes in the age-old system of land-ownership.[77] The question whether or not these reforms were appropriate in the cultural context of Afghanistan, or had any economic rationale, has evoked different responses.[78] But the fact remains that glaring inequalities and inequities existed in the Afghan countryside. The major cause was the pattern of land ownership. Although cadastral surveys of landholdings either were never made on a scientific basis or were incomplete, what statistics are available indicate that only 2.2 per cent of the landowners held 42 per cent of the land.[79] More than 400,000 peasant families or roughly 40 per cent of the peasants owned no land at all,[80] and 40 per cent of those farmers who owned land had plots smaller than 0.5 hectares. The holdings of another 40 per cent ranged from 0.6 to 3.9 hectares.[81]

The vast majority of the peasants worked as sharecroppers or as wage-labourers. Mechanization of agriculture, although at an early stage of development at the time of the revolution, was fomenting old inequities and creating new ones. Two effects of this change are worth mentioning. First was the increasing rate of unemployment among the rural poor, which in part resulted from the use of tractors. Second was the growing interest of big landowners and professional money-lenders in grabbing the farms of the subsistence farmers on the pretext of non-payment of previous loans. Better rates of return on commercialized agriculture were setting a new trend toward larger landholdings, but at the expense of small landowners. Therefore, there is no question about the necessity of land reforms for both removing inequities in the rural economy and bringing about agricultural development.[82] But the immediate goals of the Marxist regime in announcing the land reforms were more political than economic. The purpose was to establish the credibility of the new rulers as true reformers and to galvanize the support of the landless peasantry.

The land-reform law fixed a ceiling of thirty jerib (about fifteen acres) for first-grade land, or more poor land depending on its productivity. The land above this limit was to be confiscated by the government without compensation. Similarly, all lands belonging to the royal family were taken over for distribution among peasants. Six categories of 'deserving persons' were established to determine an order of preference for land entitlement. Landless peasants and agricultural workers already cultivating the land topped the priority list, while *koochis* (landless nomads) were the last category to be allotted land. Each beneficiary family was to receive the equivalent of five jeribs of first-degree land. While the land was to be distributed free, the decree set an obligation that each allottee family would cultivate the land within three months, register the deed with the government and pay a nominal land tax. These lands were non-transferable, meaning the allottees could not sell or mortgage them to obtain loans, which the peasants had traditionally done to meet their social obligations.[83]

The land-reform decree also addressed the question of water distribution, which had never been regulated on such a scale. The

Ministry of Water and Energy took control of irrigation canals and water channels. It was invested with the authority to advise the peasants on matters of cultivation. Disputes arising out of land occupancy and distribution were to be settled by the 'special committees' established at district and provincial levels. Although in theory the peasants had representation, these committees were dominated by the government and party functionaries.

The government promised the peasants benefitting from the distribution of surplus land that they would receive improved seed, fertilizer, agricultural implements and enough water to grow crops. But these were empty promises as the regime had neither the resources nor the administrative machinery to implement this programme. Cutting the peasants off from the traditional support system proved disastrous: the co-operatives envisaged under the scheme could not be formed and the much-needed loans from the Agricultural Bank were not made available. As a result there was chaos as the peasants were unable to grow anything, thus further aggravating the economic condition of the country.[84]

The relevant question to ask is, how much land did the government acquire by the decree and how much of it did it actually distribute? There are a number of difficulties which complicate an answering of this question. First, no statistics on land ownership or even on all arable land were ever compiled on scientific grounds; only rough estimates have been worked out. Second, the Marxist regime, in the best Stalinist tradition, fabricated the numbers of both land acquisitions and distribution, making it quite a complex task for any observer to find the truth. Third, official sources gave contradictory information to the public from time to time about the progress of the land reforms. Official sources claimed to have distributed 1.6 million acres among 285,000 families between January and July 1979, and to have established 964 co-operatives over 257,500 acres, which benefited 140,000 families.[85] Some of the 'independent' studies on this issue are highly biased in favour of the Marxist regime. Sadhan Mukherjee claims that 330,000 hectares of land were taken over by the Marxist government for distribution among the landless peasants, and about 1,210 agricultural co-operatives with 183,000 peasant members were formed.[86] These figures tending to show the success of the land reforms have been contradicted by other sources.[87] The scale of the uprising in the countryside and the lack of access by the government would put these figures in serious doubt. Very little, if any, land was distributed. Even in cases where it actually took place, it was not possible to implement. In August 1981 the Karmal regime made a radical revision of the land-reform legislation which entirely changed its spirit and original goal. The amendment stressed 'observing the principle of legal property, respecting the religious and tribal traditions and customs'.[88] A number of individuals and institutions were exempted from the land-reform decree. These were: a) *ulema* and the spiritual leaders, b) tribal leaders having contributed to the defence of the country, c) army officers, and d) landowners who would mechanize farming, increase production and sell their produce to the government. The new decree also promised to give lands to those Afghans who went into exile, and to compensate any individual of the above categories whose land was allotted to others. There are three reasons for the reversal of the land-reform policy. First, the reforms could not be implemented due to the collapse of the administrative machinery, and it made sense to call off the reforms through this face-saving amendment. Second, the Karmal regime, under Soviet influence, wished to cultivate a new image and to distance itself from the inappropriate policies of the Amin period. Third, Karmal might have thought that restoring land rights to the influential members of society would induce them to support his government.

Social reforms were as ambitious as the land reforms. They focused on education, with particular emphasis on adult and female education in the rural areas. The feudal and tribal social structures in Third World societies have perpetuated an irrational view of social relations between man and woman. Decree Number 7 was enacted for 'ensuring equal rights to men and women in the field of civil law', with a view to transforming the traditional status of women in Afghanistan.[89] It fixed the marriageable age at sixteen for girls, and at eighteen for boys. It limited dowry or 'bride-money' to a sum of 300 Afghanis ($10) and declared forced marriages of girls and widows illegal. According to the new law mutual consent of the couples was required and arranged marriages through parents were banned.

While this decree could be considered admirable and essentially progressive it demanded too much and too soon from a society that had lived with its customs for centuries.[90] Thus, attempts to bring about social changes through decrees proved politically counter-productive. Khalq's approach to reforming society was not only abrupt and hasty but also showed indifference to popular sentiment. The rural peasant society regarded the social reforms as atheistic in inspiration and an unwarranted intervention in family affairs. Thus implementation of the reforms without educating the peasantry provoked nationwide opposition and resistance.

FAILURE OF THE REFORMS

The Khalq regime faced many difficulties in implementing the reforms. Many of the obstacles stemmed from the political misperceptions of the Khalq leadership and its faulty analysis of Afghan society and culture. In hindsight it appears that the formulation and implementation of the reforms was more in the fashion of ideological romanticism than a pragmatic and practical approach to the pressing social and economic issues.[91] First, in articulating the reforms, Khalq violated the basic norms of Afghan political culture. Its unwise use of Marxist-Leninist symbolism not only enraged the masses but also provided a splendid opportunity for the opposition to exploit popular resentment. The social and cultural norms that had shaped and regulated individual behaviour and social relations in the villages could not easily give way to new modes of morality. The peasants' reluctance to accept the reforms was branded by the regime as reactionary, more in a mechanical and rhetorical fashion and without appreciation for the local value system. Some peasants simply refused to take possession of lands allotted to them under the reforms, saying it was not theirs. This attitude was guided more by the traditional norms than by fear of retaliation from the landlords. Similarly, social and educational reforms pertaining to the status of women, while admirable in principle, could not be imposed on an unwilling population. The change had to follow propaganda and organization and it had to be incremental. Given the traditional nature of the Afghan social system, reforms had more of a provocative than a mobilizing effect.

Second, the purges had weakened the administrative machinery of the government. The new administrative apparatus lacked skills and the absence of a serious understanding of the massive obstacles turned the reform process into an unmanageable mess. The emphasis on 'red' rather than on 'expert' led to serious difficulties. The PDPA appointees had neither a sufficient knowledge of administration nor adequate experience of political interaction with the masses.[92] The party had assumed power from the top and had never attempted to educate the masses in its revolutionary line. Third, the regime adopted harsh measures in implementing the reforms without addressing the issue of social resistance. This widened the communication gap between the regime and the masses, shifting the political mobilization in favour of the opposition. Excesses by the Khalq workers, their corruption and lack of discipline further provoked uprisings.

THE REBELLION

The Afghan masses, alienated by the reforms and repression, responded quickly to the calls for resistance from the opposition groups.[93] Despite the fact that the opposition was divided and loosely organized along ethnic and regional lines, it effectively exploited contradictions between the ideology of the PDPA regime and the popular Afghan cultural norms. The characterization of the reforms by the opposition groups as 'un-Islamic', and Soviet support for the Khalq regime, worked effectively in favour of the resistance.

While the Khalq regime opted for adventurism and relied more heavily on repression than on political communication, the resistance adopted a sophisticated political agenda for assuming leadership of the peasantry. It succeeded in popularizing its political line centred around pro-Islamic and anti-Soviet sentiments. The Marxist-Leninist symbolism of the Khalq regime and increasing repression in the name of revolutionary change with Soviet support provided sufficient credibility to the battle-cries of the opposition. The political mobilization of the Afghan masses and grass-roots support for the resistance changed the entire spectrum of the Afghan political situation. The factional strife within Khalq, alienation of the Parcham workers, the murder of Taraki and the brutalities of Amin, all presented difficulties both for the narrow-based PDPA and the Soviet leaders who had tended to believe that the task of defending and stabilizing the leftist regime would be easy.

NOTES

1 Khalq, *PDPA's Platform* (Kabul), 11 April 1966.

2 See the details of atrocities in Jeri Laber and Barnett R Rubin, *A Nation is Dying: Afghanistan Under the Soviets* 1979-87, (Evanston, IL: Northwestern University Press, 1988). Many of these atrocities were acknowledged by Babrak Karmal and his successor.

3 The Soviet leaders never treated either Khalq or Parcham or the united PDPA as developed communist organizations. None of the leaders of the PDPA was ever invited to the Soviet-sponsored conferences of the CPSU. *See, Events: News Magazine of the Middle East,* No. 45, 16 June 1978, p. 28; Louis Dupree, 'Afghanistan Under the Khalq,' *Problems of Communism* (July-August 1979), p. 39.

4 This collaboration has been frankly acknowledged by both the Saur revolutionaries and their supporters in Pakistan and elsewhere. See for instance Feroz Ahmad, 'Fauj Aur Party Ka Kirdar' ('The Role of Military and Party') in Abdullah Malik, *Yeh Mufti Yeh Mashaikh Aur Inqilab-e-Afghanistan* (The Religious Leaders and Afghanistan Revolution), (Lahore: Kausar Publishers 1986), 3rd edition, pp. 72-80. The late Afzal Bangash, the founder of the Mazdur Kisan Party of Pakistan, a Marxist-Leninist party, had this to say about the Afghan revolution: 'It was a revolution brought about by the Khalq party through military officers.' Maleeha Lodhi, 'Afzal Bangash: The man and his mission,' *The Muslim* (Islamabad) Magazine, 28 November 1986, p. 1.

5 Babrak Karmal was quoted as saying that 'Russia wanted that there should be a revolution here.' Anthony Arnold, *Afghanistan's Two-Party Communism: Parcham and Khalq* (Stanford: Hoover Institution Press, Stanford University, 1983), p. 59.

6 Selig Harrison suggests that the Shah of Iran's attempts to win Daud over through extension of aid and promise of transit route through the Gulf were meant to undermine Soviet influence. *See,* 'Shah not Kremlin touched off Afghan Coup,' *Washington Post,* 13 May 1979, p. C1.

7 *See,* first announcement of the Revolutionary Council. Radio Kabul in English,

29 April 1978, FBIS, Middle East and South Asia, Vol. 5, No. 84, 1 May 1978, pp. S1-S2.

8 *Basic Lines of the Revolutionary Duties of the Government of the Democratic Republic of Afghanistan* (Kabul: Government Printing Press, 9 May 1978).

9 *See,* Hafizullah Amin's statement in *Kabul Times* (Kabul) 23 February 1979.

10 R T Akhramovitch, *Outline History of Afghanistan After the Second World War* (Moscow: Nauka Publishing House, 1966), p. 46.

11 The 1964 constitution introduced a relatively liberal political atmosphere which allowed freedom of expression and electoral process. Leftist groups used this opportunity to publicize their views and participate in the parliamentary elections. *See,* Louis Dupree, *Afghanistan,* (Princeton; Princeton University Press, 1980), pp. 559-89.

12 Soon after the promulgation of the Press Law in July 1965, six private weekly newspapers funded by different political groups began mass circulation, These were: *Payam-i-Emroz; Khalq; Mardum; Wahdat; Afghan Mellat; Masawat. See,* Dupree, *Afghanistan,* pp. 601-2.

13 Fred Halliday, 'Revolution in Afghanistan,' New Left Review, Vol. 112, November-December, 1978, p. 19.

14 PDPA was formed on 1 January 1965 at the conclusion of a meeting convened at the residence of Nur Muhammad Taraki. This meeting is now regarded as the first party congress. The number of members attending the first congress is not known but estimates vary from 27 to 67. The programme adopted by the PDPA was directed toward achieving non-capitalistic growth, land reforms, nationalization of foreign trade, a cultural revolution, a democratic solution to national problems, a progressive foreign policy and friendly relations with the socialist countries, primarily with the Soviet Union. The founding congress considered the establishment of a national democratic government as a first step toward the building of a socialist society in Afghanistan based on scientific socialism. *See,* Sardar Shaukat Ali *Inqelab-e-Afghanistan* (Afghanistan Revolution), (Lahore: Peoples Publishing House, 1987), pp. 159-64; Anthony Arnold, *Afghanistan's Two-Party Communism: Parcham and Khalq* (Stanford: Hoover Institution Press 1983), pp. 160-77; Dastagir Panjsheri, 'A Glimpse at the History of Revolutionary Afghanistan,' *Democratic Republic of Afghanistan Annual* (Kabul), April 1978-79, p. 1433; Yuri V Gonkovsky et al., *A History of Afghanistan,* (Moscow: Progress Publishers, 1982), p. 303; Rahimullah Yusufzai, 'PDPA: Khalq-Parcham Struggle for Power,' *Regional Studies,* Vol. 2, No. 2 (Spring 1984), pp. 67-8.

15 On this point *see,* Arnold, *Afghanistan's Two Party Communism,* pp. 37-51.

16 Contrary to this Raja Anwar contends that 'not only were their basic politics identical, but their strategies were not much different.' *See,* his *The Tragedy of Afghanistan* (translated from Urdu by Khalid Hasan), (London: Verso, 1988), p. 54.

17 Dupree, *Afghanistan,* pp. 615-8. *See also,* Tahir Amin, *Afghanistan Crisis: Implications and Options for Muslim World, Iran, and Pakistan* (Islamabad: Institute of Policy Studies, 1982), p. 40

18 The Political Parties Bill was passed by both the Lower and Upper Houses of the Afghan Parliament in July 1966 and in May 1967 respectively. The king did not sign it for the next two years. In 1969, the Lower House amended the bill lowering the voting age from 25 to 20. This amendment was rejected by the Upper House. It appears with hindsight that Afghanistan's political establishment wanted to hold the first few national elections without the participation of political parties.

19 Leftist political groups in particular took their identities from their journals: *Khalq; Parcham* and *Shula-e-Javed.*

20 On the parliamentary politics of the PDPA *see,* a document 'The Establishment of the Marxist-Leninist Party in Afghanistan,' in Arnold, *Afghanistan's Two-Party Communism,* p. 172.

21 *Biography of Nur Mohammad Taraki* (Kabul: Afghan Government Publication, August 1978).

22 Hasan Kakar, 'Trends in Modern Afghan History,' in Louis Dupree et al., *Afghanistan in the 1970s* (New York: Praeger Publishers 1974), pp. 16-17.

23 Most of the left-wing parliamentarians were defeated. Only two prominent leaders—Babrak Karmal from Kabul and Hafizullah Amin from Paghman—were elected.

24 *New York Times,* 18 July 1973, p. A1

25 Muhammad R Azmi, 'Soviet Politico-Military Penetration in Afghanistan 1955 to 1979,' *Armed Forces and Society,* Vol. 12, No. 3 (Spring 1986), pp. 332-5.

26 Hannah Negaran (Pseud.), 'The Afghan Coup of April 1978: Revolution and International Security,' Orbis, Vol. 23, No. l (Spring 1979), p. 99.

27 Seventeen military officers and 1,000 troops were involved in the *coup. New York Times,* 1 August 1973, p. A13.

28 *New York Times,* 20 July 1973, p. 5

29 Hasan Kakar, 'The Fall of the Afghan Monarchy in 1973,' *International Journal of Middle Eastern Studies,* No. 9 (May 1978), p. 214.

30 Faiz Mohamad, a Parcham leader, was given the portfolio of the Ministry of the Interior.

31 See details in Kuldip Nayyar, *Report on Afghanistan* (New Delhi: Allies Publishers, 1981), p. 106.

32 Gen. Rahim Wardak, 'A Synopsis of Marxist Tragedy in Afghanistan,' *The Front: The Mujahideen Monthly,* (Peshawar), March 1978, p. 27.

33 Louis Dupree, 'Towards Representative Government in Afghanistan, part I: The First Five Steps' *AUFS Reports,* (Hanover, NH: American Universities Field Staff Reports, 1978), p.2

34 On the relationship between Daud and the left in 1975, *see,* comments of Ambassador Eliot in Thomas T Hammond, *Red Flag over Afghanistan: The Communist Coup, the Soviet Invasion and the Consequences* (Boulder: Westview Press, 1984), pp. 37-8.

35 A Rasul Amin, 'The Saur Revolution,' *WUFA (Quarterly Journal of Writers Union of Free Afghanistan,* Vol. 1, No. 2, (January-March 1986) pp. 11-12.

36 Jolanta Sierakowska-Dyndo, 'The State in Afghanistan's Political and Economic System on the Eve of the April 1978 Coup,' *Central Asian Survey,* Vol. 9, No. 4, 1990, pp. 85-97.

37 The major projects were in the Helmand and Nangarhar provinces.

38 For detailed analysis *see,* Maxwell J Fry, *The Afghan Economy: Money, Finance, and the Critical Constraints to Economic Development,* (Leiden: E J Brill, 1974), p. 47.

39 According to Anderson, commercial farmers had started taking over communal lands that were free for use by all. Money-lenders and urban land speculators also rushed to grab rich lands. This trend would have introduced a new middle class into rural society with the potential to renegotiate influence, power and status. *See,* John W Anderson, 'There Are No Khans Anymore: Economic Development and Social Change in Tribal Afghanistan,' *Middle East Journal,* Vol. 32, No. 2, 1978, p. 175.

40 Lloyd Baron, *The Paktia Development Project , Afghanistan: A History, Analysis, and Summary of Project Proposals,* (Washington, D.C: United States Agency for International Development, 1973).

41 Naum Gurevich, 'Problems of Agricultural Production in Afghanistan,' in *Afghanistan: Past and Present,* (Moscow, USSR Academy of Sciences, 1982), p. 165.

42 This is more true of those who participated in the insurgency. According to Louis Dupree, ninety-three accused insurgents were put on trial in 1975. Three of them received the death sentence. Another twenty-one were also awarded the death penalty but the President commuted their sentences to life imprisonment; two received prison term of eighteen years. The rest got from one and half years to fourteen years. Only sixteen were set free. *See,* Louis Dupree, 'Toward Representative Government in Afghanistan,' part, I, op. cit., p. 7.

43 Daud visited Pakistan in August 1976, the first visit of an Afghan head of Government to Pakistan in 20 years. *New York Times,* 22 August 1976, p. A14.

44 Harrison, 'Shah not Kremlin touched off Afghan Coup,' see note 6.

45 *Keyhan International* (Tehran), 13 June 1976.

46 'Afghanistan-Pakistan move to end long-standing feud,' *New York Times,* 27 December 1976, pp. A1, A3.

47 *New York Times,* 3 February 1975, p. A4.

48 *See,* Anthony Arnold, *Afghanistan: The Soviet Invasion in Perspective* (Stanford: Hoover Institution Press, Stanford University, 1982), p. 65; Alvin Rubenstein, *Soviet Policy Toward Turkey, Iran and Afghanistan: The Dynamics of Influence* (New York: Praeger Publishers, 1982), p. 151; Ralph H Magnus, 'Tribal Marxism: The Soviet Encounter with Afghanistan,' *Conflict,* Vol. 21, Nos. 2/3/4, 1983, p. 361.

49 This fear of Daud turning to Anwar Sadat was expressed by Babrak Karmal later in an interview. Ved Pratap Vedik, 'Interview with Babrak Karmal', *Strategic Digest,* New Delhi, 27 April-10 May 1981, p. 28.

50 Babrak Karmal mentioned the 'help of international supporters and brothers.'

See, Babrak Karmal, 'The Afghan People Will Stand Firmly on Guard of the Revolution,' *World Marxist Review (* Prague) Vol. 23, No. 4, (April 1980), p. 35.

51 Brezhnev's address to the 25th CPSU Congress in February 1976. *See, Pravda,* 25 February 1976 (*Current Digest of the Soviet Press* CDSP, Vol. 28, No. 8, 24 March 1976), p. 13.

52 *See,* R Akhramovitch's comments 'Political Shifts in the Middle East: Roots, Factors, Trends,' *World Marxist Review,* Vol. 23, No. 2 (January 1980), p. 59.

53 Feroz Ahmad, 'Afghan Developments X-rayed,' *The Marxist Review* (Calcutta), Vol. 24, No. 3 (September 1980), p. 103.

54 *Kabul Times,* 5 April 1978.

55 The central element of Daud's new approach toward Pakistan was his willingness to diffuse the issues of the Durand Line and Pushtunistan. Zulfiqar Ali Bhutto, then the Prime Minister of Pakistan, visited Afghanistan in June 1976. Daud paid a return visit in August the same year. Although *rapprochement* was only partial and had just begun to influence Kabul's posture, Daud appeared conciliatory in resolving the differences with Pakistan through direct and peaceful means. *See,* Daud's speech to the Loya Jirga, Kabul Radio in Dari, 30 June 1977, in *FBIS,* Middle East and South Asia, Vol. 5. No. 23, 3 July 1977, p. S1.

56 Brezhnev raised the issue of employment of non-Soviet experts by Daud when he visited the Soviet Union in April 1977. *See,* note 40.

57 During Daud's visit to Moscow of 12-15 April 1977 Brezhnev reportedly complained of his leaving the leftists out of his government and instead inducting objectionable (anti-communist) figures in his cabinet. The Soviet leader insisted on removing the US Aid technicians. To all this Daud is said to have responded in anger, slamming his fist on the conference table. *See,* the account in Henry S Bradsher, *Afghanistan and Soviet Union,* (Durham: Duke University Press, 1985), pp. 63-6.

58 *On the Saur Revolution (*Kabul: Political Department of the People's Armed Forces of Afghanistan, Government Printing Press, 22 May 1978), p. 11.

59 *Kabul Times,* 3 May 1978.

60 Anwar, op.cit., p. 94.

61 Louis Dupree, 'Red Flag over the Hindu Kush, Part II: The Accidental *Coup* or Taraki in Blunderland,' *AUFS Reports,* (Hanover, N. H. : American University Field Staff Reports, Asia Series, No. 45, 1979).

62 Arnold, *Afghanistan's Two-Party Communism,* p. 59.

63 Delhi Radio, domestic service in English, 27 April 1978, *FBIS,* Middle East and South Asia, Vol. 5, No. 82, 27 April 1978, p. S1.

64 Out of twenty-one ministers, only Watanyar, Qadir and Rafi were military men.

65 Radio Kabul in English, South Asia service, 30 April 1978, FBIS, Middle East and South Asia, Vol. 5, No. 84, 1 May 1978, pp. S1-S2.

66 *New York Times,* 30 April 1978, p. 10; *see,* for example reports of Taraki's press conference of 6 May 1978 in FBIS, 9 May 1978, Asia Series pp. 51-2, and in the *New York Times,* 7 May 1978, p. 1.

67 See a report by Simon Winchester, a British journalist, in *The Washington Post,* 8 May 1978, p. A16.

68 Louis Dupree, 'Red Flag Over the Hindu Kush, Part III: Rhetoric and Reforms, or Promises! Promises.' *AUFS Reports,* (Hanover, N H : American Universities Field Staff Reports, Asia series No. 232), p. 3.

69 Khalq became vociferous in claiming that Parcham had played no role in the Saur revolution, thus arguing against power-sharing on an equal basis. *See,* for instance, a controversial pamphlet published by Khalq: *On the Saur Revolution* (Kabul: Political Department of the Armed Forces, 1978).

70 In the first cabinet, formed on 1 May 1978, seven ministerial portfolios were given to Parcham and eight to Khalq. *Kabul Times,* 4 May 1978.

71 *See,* an assessment by Beverly Male, *Revolutionary Afghanistan* (London: Croom Helm, 1982), pp. 107-10.

72 It is generally speculated that the decision to send the *Parchamis* abroad was perhaps taken in the Politburo meeting of 24 June 1978. The Politburo meeting followed the Central Committee meeting of 17 June 1978 which, according to later accounts, decided that Khalq would determine state policy. Interior Minister Nur Ahmad Nur was appointed as ambassador to the US; Babrak Karmal to Czechoslovakia.

73 According to Bradsher 800 men suspected of links with Parcham were expelled from the military around the time when Karmal was removed from office. *See,* his *Afghanistan and the Soviet Union,* p. 87. The second wave of expulsions from the party and government positions started in August 1978, after the discovery of an alleged conspiracy hatched by the Defence Minister Abdul Qadir and the Army Chief of Staff Major General Shapur Ahmadzai. Both of them were arrested. They were not formal members of any organization, but were suspected of opposing radical reforms and the repressive policies of Amin and his henchmen. The Planning Minister, Sultan Ali Keshtmand, and the Public Works Minister, Mohammad Rafi, the two remaining *Parchamis,* were also arrested. Leaders of Khalq became insistent on recalling the *Parchami* leaders from abroad and instituting trials against them for their 'anti-Saur revolution and anti-Khalq regime conspiracy plotted under the leadership of Babrak Karmal'. *Kabul Times,* 24 November 1978.

74 Radio Kabul, 7 January 1980 , *FBIS,* Middle East, 9 January 1980, p. S2. For an opposite view, *see,* Beverly Male, *Revolutionary Afghanistan: An Appraisal,* (London: Croom Helm, 1982), p. 109.

75 *Kabul Times,* 13 July 1978.

76 For details, *see,* Mikhail Ilyinsky, 'The Revolution Forges Ahead,' *New Times* (Moscow)December 1980, pp. 14-15; PB Sinha, 'The Afghan Revolution and After,' *Foreign Affairs Reports,* (New Delhi), Vol. 28, No. 7, July 1979, pp. 17-18.

77 *See,* the text of Decree No. 8 in *Kabul Times,* 2 December 1978.

78 A number of scholars have argued that the land reforms were not appropriate. For this view *see,* Gerard Chaliand, *Report from Afghanistan,* (New York: Penguin Books, 1982), pp. 21-3; Anthony Hyman, *Afghanistan Under Soviet*

Domination, 1964-81, (New York: Saint Martin's Press, 1982), p. 89; Leon Wieseltier, 'Rebels with a Cause,' *The New Republic,* Vol. 3, No. 533, 4 October 1982, p. 32. For an opposite view, *see,* David Gibbs, op. cit., p. 42. Barnett R Rubin examines the reform in the context of 'redistribution' and the 'intrusive' role of the Afghan state which generated the rebellion. See his 'Redistribution and the State in Afghanistan: The Red Revolution Turns Green,' in Ali Banuazizi and Myron Weiner (eds.), *State, Elites, and Redistribution in Afghanistan, Iran and Pakistan,* (forthcoming).

79 *See,* these figures in Halliday, 'Revolution in Afghanistan,' p. 33.

80 Roman T Akhramovitch, *Outline History of Afghanistan After the Second World War,* (Moscow, 1966), pp. 8-9; Feroz Ahmad, 'Interview,' *MERP Reports,* No. 89, July-August 1980, pp. 13-14.

81 Halliday, 'Revolution in Afghanistan,' p. 33.

82 *See,* a study which was commissioned by the International Labour Organization, I Z Bhatty and L Berouti, 'A Development Strategy for Afghanistan: Lessons of an Employment Policy Mission,' *Pakistan Development Review,* Vol. 19, No. 4, 1980, p. 344.

83 *New York Times,* 13 July 1979, p. 3; Tass from Kabul, 30 July and 5 August 1979, in *FBIS*/SU, 31 July and 7 August 1979, pp. D 1 and D3-4.

84 Fred Halliday, 'Afghanistan—A Revolution Consumes Itself,' *The Nation,* No. 229 (17 November 1979), p. 493. *See also,* comments by Soviet Academician E Primakov in FBIS, *Soviet Series,* 23 January 1981, p. D1

85 *New York Times,* 13 July 1979, p. 3; *The Guardian,* 29 April 1979; Tass from Kabul, 30 July and 5 August 1979, in *FBIS/* SU, 31 July and 6 August 1979, pp. D1 and D3.

86 *See,* for instance Sadhan Mukherjee, *What is Happening in Afghanistan,* (New Delhi: Communist Party of India, Publication No. 9, July 1981), p. 19.

87 Vladimir Glukhoded, 'Economy of Independent Afghanistan,' in Social Sciences Today, Editorial Board (eds.), *Afghanistan Past and Present,* (Moscow: USSR Academy of Sciences, 1983), pp. 222-45.

88 Radio Kabul, domestic service in Pushtu, 10 August 1981. *See,* the text of the amendment issued by the Presidium of the DRA Revolutionary Council in *FBIS,* SAS, Vol. 8, No. 161, 20 August 1981. pp. C1-C2.

89 Text of Decree Number 7 was published in *Kabul Times,* 18 October 1978.

90 Nancy Hatch Dupree, 'Revolutionary Rhetoric and Afghan Women,' in M Nazif Shahrani et al., *Revolutions and Rebellions in Afghanistan* , pp. 322-4; Hannah Negaran (Pseud.), 'Afghanistan: A Marxist Regime in a Muslim Society,' *Current History,* Vol. 76, No. 446 (April 1979), p. 174.

91 Vladimir Swetozarov, 'Afghanistan: Strengthening Revolutionary Gains,' *Asia and Africa Today* (Moscow) No. 2, 1983, pp. 42-9; Richard S Newell, 'Revolution and Revolt in Afghanistan', *The World Today,* Vol. 35, No. 11 (November 1979), p. 437.

92 *See,* Karmal's assessment of the party's weakness in his 'The Afghan People will Stand Firmly on Guard of the Revolution,' op. cit., p. 34.

93 *See,* Richard F Strand, 'The Evolution of Anti-Communist Resistance in Eastern Nuristan,' in Shahrani et al., *Revolutions and Rebellion in Afghanistan,* pp. 77-93.

Chapter 3

The Soviet Intervention

The military intervention in an area outside the Soviet Union's traditional sphere of influence gave rise to a wide array of hypotheses and generalizations concerning Soviet motives in south-west Asia and the effects of her presence in Afghanistan on the geopolitical balance of the region. The active involvement of the Soviet military forces in suppressing the Afghan resistance in order to stabilize a Marxist regime and to consolidate the Soviet's strategic position along the northern fringes of the south-west Asian region revived the old apprehensions about Soviet domination in vulnerable areas on her southern periphery. The geopolitical dynamics of Soviet intervention in Afghanistan also undoubtedly altered the threat perceptions of neighbouring states.[1] It has been an almost consensual view that the Soviet Union's stable political control in Afghanistan, the success of the Sovietization process, and consolidation of a dependent Marxist regime not only would have energized the separatist and radical subnational groups in Iran and Pakistan, but also would have provided the Soviet Union with greater diplomatic leverage to neutralize the adjacent regions.[2]

Why did the Soviet Union intervene in Afghanistan? Was the intervention an extension of the Brezhnev Doctrine to non-aligned, Third World and Islamic Afghanistan? Was the Soviet invasion a calculated and well-planned move towards reaching the littoral of the Indian Ocean? What threats did the Soviet leaders perceive in the growing power of internal forces opposed to the Khalq regime? What was the nature of the interests that the Soviet Union considered vital and worthy of military intervention? What role did the geopolitical developments in south-west Asia play in shaping the Soviet perception of events in Afghanistan? These are basic questions which have formed the central focus for scholars and policy makers in various countries. While there has generally been agreement among observers on the strategic consequences of Soviet intervention and its destabilizing effects on the geopolitics of adjacent regions, their explanations of the causative factors continue to range widely from the Soviet quest for world domination to her legitimate defence of national security interests in the unstable bordering areas.[3] Before we present our views on this subject we must review the widely-debated perspectives on Soviet intervention and attempt to assess their underlying assumptions and empirical relevance to the policies of the former Soviet Union in south-west Asia during that period.

SOVIET EXPANSIONISM

Theses on Soviet expansionism rest on several assumptions about Soviet ideology, its political system and its role as a global power. One of the variants of these theses suggests that expansionism, defined loosely as ideological influence, political control and military presence, were the central dynamics of the Soviet system. Its goal of world socialist transformation, in particular through support of class wars and socialist and national liberation movements, prompted her entry into internal conflict.[4] Depending on who was at the helm of affairs in the Kremlin and the assessment of the overall correlation of forces, the Soviet Union invested heavily in struggles aiming at a socialist restructuring of post-colonial Third World societies such as Afghanistan, Angola, Ethiopia and South Yemen. This enabled Moscow to secure their adherence to its view of global politics and security.

Another perspective attempts to explain Soviet expansionism as a means of deflecting attention from domestic problems. Short victorious interventions on the pattern of Hungary (1956) and Czechoslovakia (1968) contributed to keeping the situation calm at home.[5] Belief in the greatness of Russian power, the desire to enhance the revolutionary role of the Soviet Union and overmilitarization of the system all contributed to Moscow's quest to expand its ideological, political, and security influence around the grey areas of the world within the operative framework of competitive globalism. Keeping in view our central focus on south-west Asia, we attempt to examine the relevance of expansionist perspectives in relation to a) the Soviet quest for warm-water ports and b) the continuity thesis, which argues that the geopolitical imperatives of the Soviet state were not much different from those of its predecessor.

a. The Quest for Warm Waters

The Soviet intervention in Afghanistan was seen as a well-planned and determined move to secure access to the warm-water ports in the Indian Ocean.[6] The relevance of the Indian Ocean to Soviet geopolitics goes back to the Czarist interest in extending Russian territorial domination to the northern parts of the south-west Asian region. Despite several plans to establish territorial hegemony in this area, and even contemplated co-operation with France, the Czars considered the adventure beyond the Amu Darya (Oxus River) risky and unfeasible.[7] But the fact remains that the 19th-century Russian geopolitical consciousness attached tremendous importance to the Indian Ocean for the purposes of colonial expansion.[8]

It has often been suggested that the Soviet leaders continued to entertain the old Czarist desire to expand their territorial control beyond their Central Asian frontiers to the shores of the Persian Gulf and the Indian Ocean.[9] In this connection, the most frequently quoted evidence is drawn from the Molotov-Ribbentrop talks of 1940. It is alleged that during these negotiations, the Soviet interlocuters attempted to seek Nazi acknowledgment of the Soviet sphere of influence southward in the direction of the Indian Ocean.[10]

The warm-waters theory, however, has serious weaknesses in logic and evidence. First, the historical evidence pertaining to the Molotov-Ribbentrop negotiations is controversial. The Soviet insistence on the recognition of its interests towards its southern periphery is quoted out of context. What is more probable, and perhaps accurate, is that the German side attempted to entice the Soviet Union by showing it the way to the weakening British empire in south Asia. Germany at that point was vigorously seeking

Soviet recognition of its domination in north-western Europe. In view of Moscow's vital security interests in Europe, the German 'offer' had hardly any practical value. Moreover, German influence in the northern regions of the Indian Ocean adjacent to the Soviet borders was non-existent. How then could the Nazis successfully use the region as a bargaining chip in luring the Soviet Union away from Europe? Again, with fundamental changes in the distribution of world power, new political realities and technological advancement, it is hard to see any relationship between the old colonial ambitions, if any, and the post-war strategy of the Soviet Union.

Second, the Soviet Union had emerged as a competitive naval power in the Indian Ocean[11] even without the land bridge of Afghanistan, Iran and Pakistan, although only costly modern technologies of sea power and political influence in the region facilitated the Soviet naval presence. Therefore, the Soviet move into Afghanistan was hardly motivated by its grand geo-political designs in the Indian Ocean. Its military presence in Afghanistan, though for entirely different reasons, provided the Soviet Union with the considerable advantage of geographical proximity which would have been a strategic asset had it succeeded in stabilizing Afghanistan and consolidating its military position.

b) The Continuity Thesis

The central argument of this perspective is that the Soviet intervention in Afghanistan was consistent with the pattern of Russian imperial expansion in the Caucasus and Central Asia which occurred in the latter part of the nineteenth century.[12] The conquest of Central Asia, though under the colonial impulse, was rationalized by the Czars on the grounds of 'security' and the need to 'civilize' 'barbaric' peoples on the periphery of the empire.[13] At a time when all the European powers were scrambling for territories, that was the acceptable logic of the strong against the weak. The fact is that the Central Asian Khanates were divided and vulnerable to Czarist expansionism. One after the other, they all fell under Russian control within a period of half a century. Besides their being individually small and militarily weak, two other factors contributed to their disappearance: geographical isolation and the indifference of the outside powers.

While other European nations either were preoccupied with the European balance of power or had colonial interests elsewhere, Britain was the only power that saw a threat in Russian expansionism. The British threat perception was well-calculated because the expanding Indian empire lay exposed from the direction of the Hindu Kush mountains. The British strategists recognized the importance of this traditional invasion route to South Asia. No less disturbing was the Czars' view of the British possessions in the East. Rightly or wrongly, they tended to believe that these markets and the resources of Britain's colonies in the East were the main factor which had contributed to that country's powerful role in European affairs. The Czars never concealed their jealousy and intention to subvert Britain's colonial holdings.

Britain's anxieties over the Russian expansionism in Central Asia were well-founded. The response to the Russian threat led to important political and strategic consequences for the region that continue to influence its security today. As a first step in the protection of its Indian empire, Britain decided to pursue a policy of 'forward defence' in the north-western areas adjacent to Afghanistan. But this policy, despite enormous investment of physical resources in the raising of infrastructure, could not alone ensure security without the 'neutrality' of Afghanistan, because the country was no less important to the Russians for further expansion. Obviously, both the colonial powers regarded Afghanistan as central to the security and extension of their empires. They became interlocked in a bitter rivalry over the future role of Afghanistan. Intense diplomatic manoeuvres, military preparedness, intrigues and intrusion into Afghanistan's internal affairs resulted in what is known as the 'Great Game.'[14] British policy settled on keeping Afghanistan as an autonomous buffer, but essentially under its own influence, and after failing to make any headway in its quest to establish dominance over Afghanistan, Russia acknowledged British pre-eminence. In any event, exercising influence over Afghanistan did not come easy. Britain had to fight two costly wars (1839-42, 1878-80).[15] It frequently interfered in Afghanistan's internal politics to install or protect a friendly ruler, and kept control of the country's external affairs. This prevented Afghanistan's tilt toward Russia.

Afghanistan's role as a buffer between the Soviet Union and South Asia began to erode as a result of political and strategic developments in the region in the late 1940s. First, the partition of Britain's Indian empire left the security system of the North West Frontier Province to Pakistan,[16] which had neither the global influence of a colonial power nor the resources to maintain the old balance of power. Rather, with its independence, the Pukhtunistan and Durand-Line issues that had been settled by Britain placed Afghanistan and Pakistan on a confrontational course. Second, as relations between the two worsened, the Soviet Union grasped the opportunity to woo Afghanistan by fully supporting its irredentist claims against Pakistan. Subsequently, Soviet penetration and a growing dependency relationship made Afghanistan vulnerable to Soviet designs. Observers supporting the continuity thesis see a strong relationship between the cultivation of influence by the Soviet Union and its intervention in Afghanistan.[17] Also, they have argued that while the circumstances of Soviet military action in Afghanistan and the nature of the geopolitical balance in the region at that time were different from the colonization of Central Asia in the 19th century, the motivation and consequences of intervention were essentially the same. Soviet attempts to pacify the country through varied strategies of war and political reform in the early stages of intervention strengthened the view that the Soviet Union had taken a leaf from its experience in Central Asia and was following a similar pattern of institutionalization and Sovietization in Afghanistan.[18]

The linkage pattern drawn by the continuity thesis between Soviet intervention in Afghanistan and Czarist expansion in Central Asia, however, grossly exaggerates the element of territorial expansion in the Soviet decision to invade Afghanistan by alleging that it offered an opportunity to control territories further down the Khyber Pass. The Soviet thrust into Afghanistan rather falls into the hegemonic intervention category of the Cold War era which is quite distinct from the colonial patterns of Europe or imperial Russia.

The post-World War II international system was largely shaped by the competitive and hegemonic globalism of the United States and the Soviet Union, the two emergent superpowers. The demarcation of spheres of influence in their adjacent regions and of privileged positions to define and protect geopolitical interests in such areas was tacitly recognized. The acknowledged primacy of a superpower's interests in its sphere of influence would 'entitle' it to oversee the internal political dynamics of the states under its domination in order to prevent any defection. While the spheres of influence were clearly drawn in strategically important areas of Europe and the Americas, the rest of the regions were left free for contest through diplomatic, economic and security means. What made the Soviet intervention an aberration was the fact that Afghanistan fell into those grey areas of the Third World where the Soviet Union's privileged position was never acknowledged.

The superpowers' interests did not compete in Afghanistan. The peripheral significance of the country to East-West issues rather enabled the Afghan rulers to obtain development assistance from both the blocs.[19] The US and Soviet roles as aid givers and trade partners were not directed towards outflanking each other. Although US policy makers organized a collective security system under the Baghdad pact (1956) in the 'northern tier' to ensure stability and security on the southern periphery of the Soviet Union, they left Afghanistan largely out of their geopolitical calculations.[20] The relatively greater priority given to Pakistan in the defence pact and Afghanistan's disputes with it made it difficult for the US to draw Kabul into the alliance. The US, however, provided Afghanistan with humanitarian and development assistance for decades, which did not produce the political effects that could match the growing Soviet influence.

The US did not see any interest in stepping into Britain's shoes and replaying the 'Great Game' in Afghanistan. Washington was satisfied with Afghanistan's neutral status until the Saur revolution. Even after the revolution, the US, though concerned with increasing Soviet involvement in Afghanistan, did not make any move until the Soviet Union intervened militarily.[21] The critics of US policy rightly asserted that it 'discovered' Afghanistan in December 1979. It makes sense if one considers the benign neglect which Washington showed in the seventies, when economic aid to Kabul was reduced. The US and the West were indifferent to the Soviet penetration of Afghanistan resulting from that country's dependency relationships. Either they had no idea of Afghanistan's evolving 'Finlandization',[22] or they saw no danger in it. Even so, Afghanistan was never considered to be within the Kremlin's exclusive sphere, with the right to use force to reconstruct or manage the internal political equilibrium in her favour. Moscow appeared to be satisfied with both the pro-Moscow monarchy and the republican regime—until the destabilization of the old order. The Saur revolution and emergence of anti-Soviet and anti-communist resistance gradually dragged the Soviet Union into the Afghan whirlpool, which drastically changed her perception of the stakes in Afghanistan.

THE BREZHNEV DOCTRINE

The Soviets' rationale for invading Afghanistan—helping the 'democratic forces' consolidate revolutionary gains threatened by 'foreign based subversion' and invoking their international duty—appeared somewhat similar to their intervention in Czechoslovakia.[23] Until the end of the Brezhnev era the Soviet Union vigorously pursued the defence of socialist systems threatened by internal opposition. This policy was somewhat reversed by Mikhail Gorbachev.[24] This later culminated in the granting of 'absolute freedom of choice' to the East European countries in restructuring their internal politics and economy.[25] This was in stark contrast to the Brezhnev years, when Soviet leaders had argued that the maintenance of the socialist order was the shared responsibility of all socialist countries—a reference to a concept of limited sovereignty of the socialist states under the Soviet hegemonic umbrella.

The ideological argument in defence of the socialist commonwealth against internal deviationist or adventurist dangers, both from the party and from elements doctrinally opposed to the socialist system, reinforced Soviet interest in defending its domination. In areas under Soviet domination, ideological concerns remained integral to her security imperatives. The much-publicized Brezhnev Doctrine indicated the parallelism of ideological and security interests for the Soviet Union, more specifically in Eastern Europe. The Doctrine suggested that the Soviet Union would intervene even militarily in socialist countries when their governments became unreliable or failed to ensure effective control of the state apparatus. After the ruthless suppression of the liberalization movement in Czechoslovakia in 1968 through massive use of force, Brezhnev attempted to justify the Soviet action by emphasizing that:

> when external and internal forces hostile to socialism try to turn the development of a given socialist country in the direction of restoration of the capitalist system, when a threat arises to the cause of socialism in any country—a threat to the security of the socialist commonwealth as a whole—this is no longer merely a problem for that country's people, but a common problem, the concern of all socialist parties.[26]

Soviet leaders, academicians and ideologues attempted *post facto* to rationalize military intervention in Afghanistan broadly on two grounds. First, they claimed that Moscow responded to the 'invitation'[27] from the 'legitimate' government of Afghanistan under the December 1978 treaty of peace and friendship.[28] Both the Kabul and Soviet leaders cited external threat to the stability and security of Afghanistan from neighbouring states and the 'imperialist bloc' as justification for invoking the treaty.[29] Inventing the 'invitation' and emphasizing external factors were as crude as the invasion itself. It looks superficial and devoid of any sensible reasoning to claim that Amin invited his executioners. Not a shred of evidence has been produced by Karmal or by the Kremlin establishing that Amin issued an invitation. Had he been captured alive, instead of being killed by Soviet forces, perhaps the Soviets could have succeeded in obtaining the 'invitation.' The Soviet legitimization attempts, by invoking the treaty obligations and the pretence of invitation, were merely a cover-up to blunt unexpectedly sharp criticism from Western, Third World, and Islamic countries.

The Kabul leaders in attempting to deflect questions of their own legitimacy tended to exaggerate external support for the pre-invasion uprising.[30] The fact is that a grass-roots opposition to the PDPA's ideology, reforms, and its dependence on the Soviet Union developed spontaneously and spread nationwide. This is evident from the intensity of resistance and from the protracted conflict that subsequently denied the Soviet forces effective control over large areas of Afghanistan. Undoubtedly, neighbouring states and the West supported the resistance, but their assistance assumed a significant role only after the Soviet intervention. The PDPA leaders, who never bothered to do a cold-blooded analysis of the political crisis, saw an attractive scapegoat in foreign interference. Likewise, the lack of an objective appraisal of internal political turmoil in Afghanistan, and a belief in the effectiveness of the military option, complicated the Soviet task of restoring order.

The second set of reasonings which the Soviet leaders consistently addressed was the 'internationalist duty' of the USSR to render military support to stabilize a regime which faced problems in consolidating its revolution. The denial of direct support to the regime, which presumably was on the verge of collapse under the increasing pressure of Mujahideen attacks, was considered tantamount to evasion of internationalist duty.[31] The Soviet official media rhetorically argued that 'to deny support to the Afghan revolution, to leave it face to face with the forces of imperialist reaction and aggression would have been to doom it to defeat, which would have been a serious blow to the entire communist and national liberation movement.'[32] This comment and many more like it explicitly recognize both the danger to the PDPA regime, and the Soviet commitment to defend it.

The Soviet Union's assertion of the 'external function' of its armed forces in support of 'proletarian internationalism,' though theoretically a manifestation of the Brezhnev Doctrine, was largely meant to gain support for the unpopular decision to send forces into Afghanistan. In retrospect, one can see the arbitrariness of Brezhnev and his small circle in taking the decision to intervene in Afghanistan. A top adviser to Mikhail Gorbachev, Yevgeny M Primakov, who was then also Chairman of the Council of the Union and a non-voting member of the Politburo, revealed in October 1989 that a very small group of leaders ordered the military intervention in Afghanistan.[33] As soon as Gorbachev admitted, in March 1986, that Afghanistan was a bleeding wound, selective confessions of a mistake began to emerge from many Soviet sources.[34] In October 1989 the Soviet Foreign Minister, Edvard Shevardnadze, publicly condemned the 1979 invasion of Afghanistan in an address to the Soviet Parliament. Showing strong disapproval, he described the action as 'the most serious violation of our own (Soviet) legislation, of party and civic norms, and ethical standards of the time.'[35] His claim that the decision to send troops to install and prop up a pro-Soviet regime had been taken 'behind the back' of the party clearly suggests that it was more or less a personal decision by Brezhnev. It seems he took Amin's refusal to save the life of Taraki as a personal insult. Brezhnev undoubtedly wanted not only to see the end of Amin but also to directly assist a trustworthy Karmal to capture power. This was precisely what Brezhnev and his Doctrine accomplished—at the cost of Soviet prestige and more than thirteen thousand Soviet lives[36], let alone enormous destruction in Afghanistan.

The application of the Brezhnev Doctrine to Afghanistan, a backward, Islamic and non-aligned country, was never meant to save a genuine Marxist revolution, or in any sense to promote the transition of Afghan society from tribalism to scientific socialism. Neither the appropriate social conditions nor a revolutionary apparatus existed in the country to support the tall claims of PDPA leaders. Rather, the intervention was influenced by pragmatic, conventional, political and security interests. First, the socialist credentials of a dependent elite, threatened by widespread internal opposition with Islamic orientations and hostile to Soviet influence in the country, increased Soviet stakes in Afghan politics.[37] Second, Amin, who was neither trusted nor liked by the Brezhnev group, could be made an easy scapegoat for turmoil generated by reforms and repression, and for the factionalism within the PDPA. Though he appeared to be a loner without genuine Soviet or other foreign support, he retained a strong hold among the mostly Pushtun leadership of the armed forces. He could be expected, given time, to attempt to find support from Pakistan and perhaps the United States.[38] Therefore, a skillful and surgical Soviet military strike was necessary for saving whatever was left of the revolution, and removing Amin before he could act like a 'Tito' or turn to China and Pakistan. The reason Brezhnev and his associates attached ideological significance to a purely personalized decision was to motivate the armed forces and to mobilize support within the important organs of the state and party.

The foregoing hypotheses on Soviet intervention correctly address the expansionist and aggressive consequences of the Soviet action, but they succeed only partially in explaining Soviet behaviour in south-west Asia in general and Afghanistan in particular. In our view, theses centring on expansionism in the Brezhnev doctrine of intervention have three serious shortcomings which considerably weaken their analytical power. First, while focusing on the ideological and military variables of Soviet power, these analyses tend to ignore or downgrade the internal political dynamics of Afghanistan. Second, the defensive and offensive characterizations of Soviet intervention are fuzzy and even inappropriate due to their central concern with immediate events, thus explaining very little about past Soviet influence and the relationship with Afghanistan. Finally, an overemphasis on the global aspects of Soviet ideological and strategic concerns obscures the primacy of regional geopolitics to Soviet behaviour in Afghanistan. Although there are vital links between the regional and global strategic systems, a central focus on Soviet influence in Afghanistan and the interplay of Soviet interests with the domestic politics of Afghanistan would present a more appropriate framework for understanding the context and objectives of Soviet intervention.

THE INTERNAL DYNAMICS OF AFGHANISTAN

The internal political dynamics of Afghanistan are the most appropriate reference for an analysis of the Soviet intervention. This proposition has three basic assumptions: first, that the Soviet Union attached great significance to a dependent and friendly Afghanistan within the rapidly changing geopolitical milieu of South-West Asia; second, that the political developments ensuing from the Saur revolution posed a serious challenge to Soviet influence in Afghanistan; and third, that the Soviet Union had exhausted the political means available to her, or considered them ineffective in stabilizing a pro-Moscow regime in Kabul.

The past history of Soviet penetration of Afghanistan and their geographical proximity played an important role in shaping Soviet perceptions of Afghanistan's relevance to their geopolitical interests in the adjacent regions. To preserve Afghanistan as a dependent and client socialist state against the rising power of the anti-Soviet forces was perceived by Moscow to be vital to its future role in south-west Asia.[39] The apparatus of the Afghan state, with ideological ties to the Soviet Union and supported by it, had begun to crack everywhere, as a reaction both to the repression of the Khalq regime and to the Soviet influence on its leadership and policies. As the factional quarrels within the PDPA intensified, the Soviet leaders started drifting towards a more active role in Afghanistan's politics and security.

In the domestic political context of Afghanistan, four factors played an important role in influencing Moscow's decision to send its forces into Afghanistan. First, the Soviet leadership had become convinced that Amin and his close associates had no intention of promoting reconciliation with the Parcham leadership. Infighting among various factions based on ethnicity and personal loyalties had considerably weakened the party and undermined the stability of the pro-Moscow socialist elite. Second, the Soviet advisers stationed in Afghanistan became increasingly apprehensive of the growing strength and massive popularity of the resistance groups among the Afghan public. They feared that any further weakening of the regime could bring into power the Islamic fundamentalists. Third, the siege mentality of the Amin regime, doubts about Amin's loyalty, and his inappropriate socialist programme all had a significant impact on the Soviet decision to replace him by installing the relatively pragmatic and politically reliable Babrak Karmal. Finally, the balance of power in South-West Asia underwent a fundamental change with the collapse of US security arrangements in Iran. The Soviets did not fear any counter-move by the Western bloc, and the regional states lacked the required level of capabilities to deter the Soviet move. The Soviet leaders carefully evaluated both the risks and the opportunities before they decided to invade

Afghanistan. We would like to examine these and other influences on Soviet behaviour in some detail in order to explain the links between the internal dynamics of Afghanistan and Soviet perceptions of their role and their decision to intervene.

a) Disintegration of the PDPA

The factional collaboration of Khalq and Parcham under the reunified and reactivated PDPA was essentially expedient. Leaders of both groups feared, and rightly so, that their default on evolving a common political strategy would not only tighten Sardar Daud's control over the Afghan polity but would also end their political careers. They interpreted Daud's tilt toward the Islamic bloc, which he had started by seeking *rapprochement* with Pakistan, Iran and Saudi Arabia in 1976-7, as directed towards terminating leftist politics. The confrontation between Daud and the socialists had been simmering since the purging of the *Parchamis* from the government. Increasing intolerance of the Daud regime prompted reunification of the Marxist groups.

The 1978 *coup,* which put an end to the monarchy dominated by the Muhammadzai lineage of Durrani Pushtuns, brought into power a wide mix of middle-class intelligentsia, army officers and members of influential families unrepresented in the previous regimes. These men with a narrow support base had very little in common except their opposition to Daud and to the ideological romantics. At the first contact with Afghan political reality, the clash of personalities and particularistic interests began to erode their unity. The Afghan revolutionaries showed hardly any interest either in evolving optimal consensus on pressing national issues or in building support among the masses. Not surprisingly, within only months of the revolution the intensive power struggle between Khalq and Parcham began to affect the political capacity of a party that had limited cadres and enjoyed very little popular support.

The Soviet advisers in Afghanistan came to believe that Amin was the real culprit behind the factional bitterness. By managing the ouster of Parcham he assumed all powers, thus reducing Taraki to a figurehead. He was also considered responsible for the ruthless implementation of reforms which provoked so much hostile reaction against the Saur revolution. In the Soviet assessment, a compromise between Taraki and Karmal by neutralizing Amin could prevent further deterioration of the situation. Perhaps this was the last option available to Moscow short of direct military intervention. Sending Karmal to Prague as ambassador worked to his advantage by affording him frequent opportunities for discussing with the Soviet ambassador and other leaders of the CPSU his future role in Afghan politics.

The Soviet leaders reportedly arranged a meeting in September 1979 between Karmal and Taraki when the latter visited Moscow briefly on his way back from Cuba. Taraki succumbed to Soviet pressure and agreed to get rid of Amin, a confidant who had worked with him for a long time.[40] It would be pertinent to mention that Taraki and Amin had developed some differences, particularly on retaining some of the Parcham sympathizers in the government. The growing power of Amin, which he acquired by appointing close relatives and friends to high positions, worried Taraki. Needless to say, Taraki all along had the feeling of having been reduced to titular head of the party and the government.

The 'Moscow plan' to remove Amin did not work. Amin was informed of the Taraki-Karmal meeting by his close confidant Jadon Tarun, who was a member of Taraki's entourage. Amin wasted no time in taking precautionary measures and alerting his men in the armed forces. The encounter, of which there is hardly any authentic account, resulted in the removal of Taraki and his later secret execution on 9 October 1979.[41] Brezhnev and the Soviet advisers in Afghanistan attempted to save his life but failed. Every time the Soviets approached Amin to spare Taraki, they were told that he was safe. Amin brazenly bluffed. The Soviet Embassy, however, succeeded in giving protection to some of Amin's opponents who were later smuggled out of the country.[42] Bad blood had been created between Amin and the Soviet leaders, and the Soviets knew that the 'Moscow plan' prepared by them was an open secret to him.

The tragic end of Taraki put an end to the political option of reunifying the PDPA under him and helping the party emerge as a mass organization. Now, nothing short of Soviet intervention or total withdrawal of support could dislodge Amin from power. Withdrawal of support from him, however, was not an option as not only would it dismantle whatever was left of the socialist apparatus but it could also bring into power forces hostile to the Soviet Union.

b) Distrust of Amin

The leadership style and nationalist orientation of Amin's 'communism' were regarded by the Soviet leaders as serious obstacles in their efforts to guide and direct the domestic reform policies of Afghanistan.[44] More than Amin's nationalism however, it was his divisive role in Khalq-Parcham politics that caused concern in Moscow. For one thing, the professing of 'independence' by Amin did not accord at all with the near-absolute dependence of his regime on the Soviet Union for economic and military aid.

What worried the Soviet Union, and friends of the Afghan revolution elsewhere, was an element of despotism in Amin, and his repression of political opponents. Even Taraki, once acclaimed by Amin as a 'great teacher,' fell victim to it. Amin's gradual control of government institutions including the armed forces isolated Parcham and severely restricted Taraki's actual exercise of power. Amin's autocratic centralism led to dangerous consequences for party unity, further reducing the ability of the regime to consolidate. Furthermore, whatever balance and moderation had existed in the ranks of the party and its leadership vanished with the concentration of power in the Amin group.

If the objectives were to pull the socialist groups together and combat the Mujahideen resistance effectively, then Amin was not the right man in Moscow's estimate. The Soviet leaders had come to believe, as was evident from the 'Moscow plan', that the removal of Amin could be the first step towards paving the way to party unity and broader reconciliation within the country. Ostensibly, the Soviet leaders accepted the removal and murder of Taraki as a *fait accompli* and pledged to extend support to Amin's embattled regime against internal resistance. The Soviets kept reassuring Amin through words and deeds that their economic and military assistance to Afghanistan would not be affected by Taraki's murder. However, compared to the usual Soviet greeting sent to the most favoured socialist leaders, the congratulatory messages conveyed to Amin by Brezhnev and Kosygin were cold and smacked of indifference.[45]

Although Amin kept acknowledging the vital role of the economic and military aid extended by the Soviet Union for socialist reconstruction and counterinsurgency operations against the Mujahideen,[46] he was not unmindful of the Soviet attempts to oust him from power. A trusted lieutenant of Amin, Foreign Minister Shah Wali, at a meeting of ambassadors from the Eastern bloc, accused Alexander Puzanov, the Soviet ambassador to Afghanistan, of involvement in the conspiracy to remove Amin. The Soviet ambassador, who was represented by Vasil Safronchuck at the meeting, was charged with giving protection in the Embassy to Amin's opponents from Parcham who were involved in the conspiracy to kill him.[47] The accusations of Puzanov's meddling in Afghan affairs were followed by a request to the Soviet Union for his replacement.[48] Accusing the Soviet ambassador of interference in domestic politics was in a way paradoxical. Amin along with other party members had not only accepted 'interference' but regularly invited Soviet diplomats and advisers to mediate their disputes. Amin resented Puzanov's conduct only because the ambassador had in his judgment lost impartiality. A wiser Amin would have known that the ambassador could not and did not act independently.

In choosing to ignore the Soviet requests to spare Taraki's life when he was still alive in captivity, Amin must have thought that the event would eventually lose its significance. He may have assumed that he would be adopted by the Soviet leaders, thinking that Moscow would accept the reality on the ground. He certainly faltered in gauging the Soviet reaction and options. Neither his expressions of gratitude to the Soviet Union nor his professions of loyalty to the socialist Commonwealth earned him Brezhnev's affection. He had hardly any alternative in a situation of almost total dependence on the Soviet Union, when even the sensitive ministries were staffed by Soviet advisers.[49] Practically, he was isolated, and his hands were tied. He was entirely at the mercy of the Soviet leaders.

The Amin regime faced other difficulties. Mounting resistance against the misconceived radicalism of Khalq workers paralyzed the administration in the country. Repression unleashed by the inexperienced urban-based Marxist cadres forced thousands of Afghan nationals to flee their country and they began seeking refuge in neighbouring Iran and Pakistan. The persecution of Parcham members, one-time coalition partners, considerably alienated a sizeable number of the Afghan leftists who had worked for and supported the Saur revolution. Taraki's supporters and the disillusioned workers of all leftist factions turned hostile to Amin for not pursuing a united-front strategy. The Soviet leaders had all along viewed Amin's policies concerning both the consolidation of revolution and the pacification of resistance as misdirected. The desertion of Afghan army personnel in large numbers posed another problem. The shrinking ranks of the army severely limited the capacity of the Amin regime to calm the widespread turmoil. Needless to say, Amin lacked the credibility and the political capacity to take the initiative towards resolving the complex problems that his regime had brought upon itself.

The Amin regime had became helplessly weakened by October 1979. At this point the question was, who would replace him, the Mujahideen or the Soviets? The Mujahideen, riding on widespread unrest, posed a serious threat, but they lacked proper organization, unity and acceptable leadership and did not present a real alternative to the fast-disintegrating Amin government. The Soviet Union, on the other hand, had been planning carefully for intervention. General I G Pavlovskiy, the Deputy Defence Minister and Commander of the Soviet ground forces, had reported back to the Kremlin, on the basis of his military mission into Afghanistan, that the collapse of the Amin regime was imminent.[50]

Having got wind of the Soviet intervention plans, or at least of the web of conspiracies to remove him, Amin made attempts to reach out to Pakistan.[51] Immediately after discovering the 'Moscow plan' and Taraki's arrest, Amin started sending signals to Pakistan seeking to restore contact and to 'eliminate misunderstandings.[52] These messages to Pakistan were in fact desperate calls for help in influencing the situation inside Afghanistan. When Pakistan agreed to send Agha Shahi, then Foreign Minister, in December, it was already too late. The schedule for Shahi's visit almost coincided with the crossing of Soviet troops in northern Afghanistan. His plane could not fly to Kabul due to 'bad weather.' Pakistan was in any case in no position to defer the Soviet intervention. With advance knowledge of the impending invasion, Pakistan could at best do some drum-beating, which it did. The Soviet Union had carefully assessed the limitations of the US and the West in terms of both their regional diplomacy and the 'Pol Pot syndrome' of the Amin regime. Even if Amin wanted to defect to China, Pakistan or the West, given his siege mentality he had no time and little chance of success.

No less threatening to the Soviet Union was the prospect of Mujahideen overrunning the Amin regime.[53] But, more than a 'friendly' intervention to deny the opportunity to a resistance that was hostile to the Soviet Union, it was, undoubtedly, an invasion to remove a dependent ruler who had fallen from grace. Unlike the previous *coup* attempts, Soviet armed personnel actively participated in the killing of Amin and the installing of Babrak Karmal.[54]

c) The Rise of Mujahideen Resistance

Among the various internal political dynamics of Afghanistan that shaped the Soviet decision, the spontaneous rise of Mujahideen resistance with grass-roots support was seen as the most potent threat to the socialist regime and the Soviet Union's influence in that country. The disintegration of the PDPA coalition and the failure of the Amin regime were, in part, both cause and consequence of the activities of the Afghan resistance.

The violent opposition to the regime and its reform process put up by influential sections of Afghan society presented the opposition groups with unrestricted access to the population for provoking revolts. As the struggle progressed, Afghan nationalism and Islam emerged as the mobilizing ideologies of the resistance. All resistance leaders, though with very diverse social backgrounds, began to focus on nationalism and Islam, which in Afghan history and tradition have remained inseparable. The popular appeal of Islam and nationalism became identified with two struggles. The first was the struggle to topple a Marxist regime which was repugnant to the traditional Afghan conception of Islamic ideology. The second was the struggle against Soviet influence, which was perceived as having been instrumental in the Saur revolution and in the protection of the Khalq regime.

The Soviets saw the Mujahideen threat more in terms of losing their influence in Afghanistan than of fearing any spill-over effects of their possible victory in the adjacent Central Asian Republics. The emergence of an anti-Soviet Islamic regime in Afghanistan would have two serious implications for Moscow. First, it could undermine the credibility of the Soviet protective umbrella that was extended to a variety of Marxist regimes and revolutionary movements. Impressed by the revolutionary phraseology of the Afghans and having declared their revolution as progressive and democratic,[55] the Soviets upgraded the level of support to the Khalq regime. Moscow blundered in viewing the uprising in a mechanical fashion, regarding it as the usual reaction of the feudal and 'reactionary' classes fighting to protect class interests.[56] The increasing threats to the Khalq regime, therefore, correspondingly enhanced the level of Soviet involvement. Largely due to its overcommitment to the defence of what it came to believe to be the new order, the Soviet Union confronted a dilemma in the impending disintegration of the Khalq regime under Amin. The engulfing crisis in Afghanistan put the credibility and prestige of the Soviet Union to the test. Having encouraged and supported the revolution, Moscow could not tolerate the prospect of it being overturned by the Islamic resistance.

Second, Soviet politics in Afghanistan in the seventies had become interventionist and might be characterized as a small-scale replica of British paramountcy. Soviet geopolitical interest in Afghanistan had grown in parallel to its penetration. The degree of influence that it had gained, and the latent aspiration to seek political space towards the south-west Asian region, strengthened the Soviets' role perception in Afghanistan.[57] The first task in carving out a continental sphere of interest across the Amu Darya was to prevent Afghanistan from falling into the hands of anti-Soviet resistance groups. Inaction and default could only lead to the loss of political gains in Afghanistan. Direct involvement in the internal affairs of the country and, as a last resort, military action were opted for by the Soviet leaders as the means to exercise control over events in Afghanistan.

THE REGIONAL SCENE

The Soviet intervention in Afghanistan cannot be explained fully without taking into account the changing power equilibrium in the region toward the late seventies. The emergent trends in the power equations in the south and south-west Asian regions favoured the Soviet power projection. US and Western interest in the regional security systems had considerably weakened. Post-Vietnam US diplomacy, fashioned within the general framework of *rapprochement* with the Soviet Union, emphasised selective commitments. It later encouraged and supported regional powers to assume their security responsibilities. While maintaining a less visible role behind the scenes, the Nixon Doctrine focused on enhancing the military capabilities of friendly regional powers in strategically important areas of the world.

The Shah of Iran assumed the role of regional *gendarme* in south-west Asia under the indirect protective umbrella of the United States. The Shah's ambitions to develop Iran into a regional military power coincided with the US desire to limit foreign involvements. The United States had also begun to modernize the armed forces of Saudi Arabia. The petro-dollars in both these countries could buy the needed equipment in the US and Western Europe. The 'twin-pillar' policy had enormous advantages for the US security preferences in this region. It not only helped recycle the petro-dollars but also significantly reduced the US economic burden of foreign security assistance. But this policy failed to envisage the destabilizing potential of revolutionary movements and the inability of rulers such as the Shah to cope with a widespread uprising. Tied to the personal power of the Shah, the security arrangements collapsed with the fall of the monarchy. The United States confronted the uncertain situation with no viable alternative for influencing events.

The regional crisis resulting from the Iranian revolution coincided with the intensification of civil war in Afghanistan and the political isolation of the military regime in Pakistan which had hanged Zulfikar Ali Bhutto. The collapse of the monarchy in Iran and the pre-occupation of the United States with the outcome of the Iranian revolution had considerably reduced the possibility of an effective response to Soviet moves in Afghanistan. The situation in the region greatly influenced Soviet perceptions of a low-risk and quick victory through military intervention.

The Soviets' perception of *detente* and their assessment of the 'correlation of forces' had a significant impact on their view of events in South-West Asia and developments in Afghanistan. By the time they decided to intervene in Afghanistan the *detente* process between the superpowers was under serious strain. At this point, the Soviets believed that the SALT accord would be rejected by the US Senate. The Carter Administration had decided to increase the defence outlays and deploy new missiles in Europe. Moreover, *detente* from the Soviet point of view did not mean the freezing of the international status quo,[58] particularly in the areas adjacent to the USSR. The Soviet Union's continuing support for the revolutionary process and for wars of national liberation was seen by its leaders as entirely compatible with peaceful coexistence.[59] *Detente* with the West was even regarded by the Soviets as creating favourable conditions for the peaceful socialist and communist transformation of the world.[60] If the original American hope had been that *detente* would effectively lead the Soviet Union to forego opportunities or to willingly lose accumulated security interests through inaction, then it was misplaced. Events in Angola, Ethiopia, South Yemen and Afghanistan had sufficiently demonstrated the fallacy of unrealizable expectations attaching to *detente* during the Brezhnev era.

The Soviet leaders had never altered their bipolar world view and, though strategic nuclear parity had brought realism to the Soviet Union's earlier rhetoric on changing the entire world in its image, in the revolutionary conflicts in Third World areas, where the United States' ability to manage security was considered weak, the Soviet leaders encouraged radical groups to gain power and stabilize their control. The development of military technologies and their application, which helped to enhance Soviet power, also influenced Moscow's decisions to undertake military missions in Asia and Africa. The 'external function' of the Soviet armed forces was to defend the socialist commonwealth as well as to assist in the success of 'national liberation' movements which gained significance in areas where the likelihood of confrontation with the United States was low or nonexistent.[61] And, of course, Afghanistan was a country which was seen by the West as irrelevant to its security needs and as isolated from her neighbours because of the manipulation of border and political disputes by Moscow. The Soviets perceived greater risks in letting Afghanistan take its natural course in the intensified civil war than in acting decisively to stabilize Moscow's power and influence.

NOTES

1 Sultan Muhammed Khan, 'Pakistani Geopolitics: The Diplomatic Perspective,' *International Security,* Vol. 5, No. 1, Summer 1980, p. 26.

2 *See,* Selig S Harrison, *In Afghanistan's Shadow: Baluch Nationalism and Soviet Temptations,* (New York: Carnegie Endowment for International Peace, 1981).

3 On the Soviet point of view, *see,* Gulam Muradov, 'National Democratic Revolution in Afghanistan: A Soviet View,' *Journal of South Asian and Middle Eastern Studies,* Vol. 6, No. 1, Fall 1982, pp. 57-74; A Djabarov, 'Afghanistan: The Revolutionary Process is Irreversible,' *Soviet Military Review,* March 1981, pp. 49-51.

4 Richard Pipes, *US-Soviet Relations in the Era of Detente* (Boulder: Westview Press, 1971), pp. 135-214; *Survival is not Enough: Soviet Realities and America's Future* (New York: Simon and Schuster 1984), pp. 37-110. *See also,* a discussion by Richard Pipes and George Kennan in 'How Real is the Soviet Threat?' *U.S. News and World Report,* 10 March 1980, p. 33.

5 On this perspective, *see,* Adam B Ulam, *Expansion and Coexistence: The Soviet Foreign Policy 1917-73,* 2nd edition (New York: Holt, Rinehart and Winston, 1974); *The Rivals: America and Russia since World War II* (New York: Viking Press, 1971); *Dangerous Relations: The Soviet Union in World Politics* (New York: Oxford University Press, 1983).

6 *See,* President Carter's *State of the Union Message to the United States Congress* (Islamabad: U.S. International Communication Agency, 1980), p. 6.

7 *See,* Helene Carrere d' Encausse, 'Systematic Conquest, 1865 to 1884: Organizing and Colonizing the Conquered Territories,' in Edward Allworth (ed.), *Central Asia: Bukhara and Khiva, 1865-1924* (Cambridge: Harvard University Press, 1968); Olaf Caroe, *Soviet Empire: The Turks of Central Asia and Stalinism,* 2nd edition (New York: St. Martin's Press, 1967).

8 See details in Rasul B Rais, *The Indian Ocean and the Superpowers. Economic, Political and Strategic Perspectives* (Ottowa, NJ: Barnes and Noble Books, 1987), pp. 104-11.

9 Sir Cyril Packard, 'Afghanistan II: Difficult Decisions for the West,' *The Round Table,* No. 278, April 1980, p. 132.

10 R J Sontag and J S Beddie (eds.), *Nazi-Soviet Relations 1939-1941: The Documents from the Archives of German Foreign Office* (Washington, D.C.: Department of State, 1948), p. 257.

11 Michael McGwire, 'The Rationale for the Development of Soviet Seapower,' *Proceedings,* No. 105, May 1980, pp. 154-83; Elmo R Zumwalt, Jr., 'Gorshkov and His Navy,' *Orbis,* 24 (Fall 1980), pp. 491-510; Rasul B Rais, *The Indian Ocean and the Superpowers,* pp. 104-53.

12 On this perspective, *see,* A M Manzar, *Red Clouds Over Afghanistan* (Islamabad: Institute of Policy Studies 1980); Fath-ur-Rahman and Bashir A Qureshi, *Afghans Meet Soviet Challenge* (Peshawar: Institute of Regional Studies, 1986); Packard, op. cit.

13 Prince Gorchakov's circular memorandum to the Russian Foreign Service, 9 November 1864. A E Seen (ed.), *Readings in Russian Political and Diplomatic History,* Vol. 1 (Homewood, Ill.: The Dorsey Press, 1966), p. 140.

14 David Fromkin, 'The Great Game in Asia,' *Foreign Affairs,* Vol. 58, No. 4 (Spring 1980), pp. 936-51.

15 Captain Henry Havelock, *Narrative of the War in Afghanistan in 1838-9* (Henry Colburn Publisher 1840); William Trousdale (ed.), *War in Afghanistan, 1879-80: The Personal Diary of Major General Sir Charles Metcalfe MacGregor* (Detroit: Wayne State University Press 1985).

16 Sir Olaf Caroe, 'Afghanistan: The After-Effects of Partition,' *The Round Table,* No. 278, April 1980, pp. 126-42.

17 Tahir Amin, op. cit., pp. 45-66.

18 S B Majrooh and S M Y Elmi, *The Sovietization of Afghanistan* (Peshawar: Afghan Information Centre, 1986).

19 Nake M Kamrani, 'Afghanistan Under Soviet Occupation,' *Current History,* May 1982, p. 219.

20 'Who Lost Afghanistan?' *Time,* 28 January 1980, p. 16; *See also,* Mohammad Islam, 'US-Afghanistan Relations: A Historical Perspective,' *Pakistan Journal of American Studies,* Vol. 9, Nos. 1&2, Spring/Fall 1991.

21 As early as 20 September 1979, the State Department had reported increased activity of Soviet armed forces near the northern borders of Afghanistan, and warned the Soviets against intervention. *International Herald Tribune,* 21 September 1979.

22 'What Lies Broken in Afghanistan?' (editorial), *New York Times,* 9 January 1980.

23 According to Jiri Valenta, the story of Czechoslovakia repeated itself in a more daring fashion. See his 'From Prague to Kabul: The Soviet Style of Invasion,' *International Security,* Vol. 5, No. 2, p. 34.

24 Gorbachev, in a meeting with Georges Marchais, said, 'The principles governing relations among socialist countries and among communist parties have been defined clearly and for all time: full autonomy and full independence, no interference of any kind.' *Pravda,* 23 September 1989, p.1 (CDSP, Vol. XLI, No. 38, 18 October 1989, pp. 23-4.

25 Edvard Shevardnadze's speech before the Soviet Parliament on 23 October 1989. *The Economist,* 28 October 1989, pp. 15-16.

26 *Pravda,* 13 November 1968, pp. 1-2, CDSP, Vol. 20, No. 46, 4 December 1968, pp. 3-5.

27 On Soviet official justification of the invasion in the Brezhnev period, *see,* Gulam Muradov, 'National Democratic Revolution in Afghanistan: A Soviet View,' op. cit. pp. 57-74; Yuri Gankovsky, 'International Relations, Peace and Security: Problem of South Asia in the Works of Soviet Orientalists,' *Strategic Studies,* Vol. 6, Nos. 2 and 3, Winter/Spring 1982-83, pp. 21-7; A Djabarov, 'Afghanistan: The Revolutionary Process is Irreversible,' *Soviet Military Review,* March 1981, pp. 49-51.

28 The Treaty of Peace and Friendship was signed on 5 December 1978 during Taraki's visit to Moscow. For the text *see, Kabul Times,* 9 December 1978.

29 *New York Times,* 4 January 1980, p. 1.

30 In less than a month after the revolution, rebellions broke out in Nuristan on 22 May 1978 and again on a larger scale on 4 October 1978. Insurrections in the Hazarajat, Laghman and Kunar Valley were also reported.

31 'World Communist Solidarity with the Afghan Revolution,' *New Times,* No. 3, January 1980, pp. 9-10.

32 Ibid.

33 *See,* a statement of the President of the Council of Soviets and Non-Voting Member of the Politburo, *New York Times,* 28 October 1989, p. 6.

34 In an interview Marshal Sergei Akhromeyev, Gorbachev's top military adviser, said, 'It (the intervention) was certainly the wrong decision.' *Time,* 13 November 1986, p. 59.

35 *Washington Post,* 24 October 1989, pp. A1, A22.

36 The Soviets first released figures for losses in Afghanistan on a yearly basis in August 1989. *Pravda,* 17 August 1989, p. 6; CDSP, Vol. XLI, No. 33, 13 September 1989, pp. 32-3.

37 'Afghan Action Accompanied by Extension of Soviet Intervention Doctrine,' *Soviet World Outlook,* Vol. 5, No. l, 15 January 1980, p. 2.

38 The Soviets alleged that Amin had held confidential meetings with American officials and sent personal oral messages to President Jimmy Carter. *Far Eastern Economic Review,* 7 March 1980, p. 15.

39 *See,* Vladimir Petrov, 'New Dimensions of Soviet Foreign Policy,' in *Evolving Strategic Realities: Implications for U.S. Policymakers* (Washington, D.C., National Defense University Press, 1980), p. 33.

40 Author's interviews with exiled Afghan leaders in Peshawar, Pakistan in January 1990.

41 *Kabul Times,* 10 October 1979.

42 Sarwari, Qadir and Watanyar were the prominent leaders sought by Amin who were given refuge by the Soviet Embassy.

43 Contrary to this, Anwar contends that Amin trusted the Soviets till the last day of

his life and never suspected what was in store for him. *See*, his *The Afghan Tragedy*, pp. 182-93. In our view this position is neither logical nor tenable.

44 Beverley Male, op. cit., pp. 156-70.

45 Brezhnev's greetings to Amin read: 'Congratulations on your election to the post of General Secretary of the Central Committee of the People's Democratic Party of Afghanistan, Chairman of the Revolutionary Council and Prime Minister of DRA.' *Pravda*, 17 September 1979. The greetings to Babrak Karmal read: 'I cordially congratulate you.....on behalf of the Soviet leadership and on my own behalf, I wish you great success in all your multifaceted activities for the good of the friendly Afghan people.' *Pravda*, 29 December 1979.

46 See one of his routine statements expressing gratitude to the Soviet Union. Radio Kabul, 7 November 1979, in FBIS/ME 9 November 1979, pp. 51-3.

47 *The Economist*, 3 November 1979, pp. 52-3.

48 FBIS/ME, 9 September 1979, p. 56.

49 'A Chronology of Soviet-Afghan Relations: April 1978-January 1980,' *Radio Liberty Research*, RL 17/80, 2 January 1980; *see also, The Economist*, 17 November 1979, pp. 68-9.

50 General I G Pavlovsky visited Afghanistan for nearly two months, from late August 1979 to October 1979, to study the situation from the perspective of military contingency. *See*, reports in *New York Times Magazine*, 13 January 1980, p. 37; *The Wall Street Journal*, 24 April 1980; *Newsweek*, 14 January 1980, p. 10.

51 'Amin reportedly appealed to Zia,' *Washington Post*, 14 February 1980.

52 Radio Kabul, 17 September 1979, in FBIS/ME, 18 September 1979, pp. 51-5; *see also*, a report in the *Indian Express* (New Delhi), 13 February 1980, p. 6.

53 *New York Times*, 3 January 1980, p. 1.

54 *See*, the report of a Pakistani diplomat who was posted in Kabul at the time of the invasion. Altaf A Shaikh, 'Soviet Intervention in Afghanistan: The Assassination of Hafizullah Amin,' *The Nation* (Lahore), 24 August 1988.

55 In Baku, Azerbaijan, Brezhnev talked of 'people's revolution in Afghanistan,' toppling of 'semi-feudal regime,' and 'new Afghanistan.' *Izvestia*, 23 September 1978, pp. 1-2, CDSP, Vol. 30, No. 38, 18 October 1978, pp. 1-4.

56 Dmitry Volsky, 'The Target: Afghanistan's Revolution,' *New Times*, No. 24, 1979, pp. 12-13.

57 In an interview with a West German journalist the late Yuri Andropov drew a parallel between US interests in Central America and Soviet interests in Afghanistan. *Dawn* (Karachi), 25 April 1984.

58 Brezhnev's speech before the 25th Party Congress. *Pravda*, 25 February 1976, pp. 2-9, CDSP, Vol. 28, No. 8, 24 March 1976, pp. 3-33.

59 Vikenti Matveyev, 'The Soviet Viewpoint,' *Africa Report*, Vol. 21, No. l, January - February 1976, p. 67.

60 *Izvestia*, 25 February 1976, p. 14.

61 A Korkeshkin, 'Strength of the Soviet Army,' *Soviet Military Review*, No. ll, November 1972, p. 6; Sergey Gorshkov, *The Seapower of the State* (Moscow: Military Publishing House, 1976), pp. 2, 268-9.

Chapter 4

The Soviet-Afghan War

A quick look at the composition and structure of the interventionist forces would reveal that the Soviets had two immediate objectives. The first was to pre-empt and effectively fight any resistance from the Afghan armed forces. It was to meet this contingency that the Soviet forces entered Afghanistan equipped with man-portable anti-aircraft missiles and anti-tank weapons.[1] Second, the most urgent task assigned to the forces was to remove Hafizullah Amin and install Babrak Karmal. Apparently while the original Soviet objectives were limited, Moscow's strategy of political control, transformation of the regime and stability had its own momentum. In the process it became neither a holding operation nor merely a deterrent posture in support of the beleaguered Kabul forces. The Soviet military apparatus became increasingly involved in securing strategic areas, launching punitive strikes and undertaking pacification campaigns.

The Soviet military planners made serious miscalculations in assessing the political and military situation in Afghanistan.[2] First, they thought that the new government headed by Karmal could and would control the situation. Second, they underestimated the widespread and popular nature of the uprising against the Kabul regime. Third, they misjudged international reaction and the ensuing foreign military support to the Afghan resistance. Finally, they expected a quick military victory over the scattered and unorganized rural-based resistance.[3]

The Soviet failure to make an objective analysis of situations such as Afghanistan was often rooted in the structure of its policy-system, which tended to overlook internal disaffection and shifted the blame on to the foreign powers. All Soviet leaders consistently attempted to legitimize their invasion of neighbouring countries by linking foreign support with the forces they intended to defeat.[4] They believed, or at least pretended to believe, that the local population supported the beleaguered regime and that the armed opposition could be separated from the masses and eventually destroyed. Such a poor understanding of the political crisis led the Soviet Union to apply military tools to an essentially political war; the use of force in reconstituting a 'legitimate' political order in effect deepened the crisis. Oblivious to the objective realities of Afghanistan, the Soviet forces entered the country psychologically and materially unprepared to fight a war against the entire Afghan nation, which they had not expected to last so long.

The Soviet strategists shaped their doctrines of intervention and counter-insurgency in the light of their experience of fighting internal wars for consolidating the Bolshevik revolution, and of successful military operations in Eastern Europe.[5] In these conflicts they had learnt the value of surprise, speed and the application of overwhelming force for making foreign military campaigns effective. In each situation, the Soviet Union rapidly moved a combination of air and land forces in conventional-sized configurations to seize the major command and control targets and then moved out to defeat the insurgents beyond the defensive perimeters. The Soviet counter-insurgency tactics in Afghanistan were also an extension of its operations in Central Asia against the Mujahideen, generally characterized as *basmachi* (bandits) by the Soviet propaganda machine.

The Soviet counter-insurgency doctrine comprised three elements. First, it focused on the creation and strengthening of a mass-based political organization. The Soviets used ideological symbols to attract some of the alienated groups of the subject societies. They co-opted local figures into political roles. They had found that the promise of social mobility and incentives of economic opportunity induced many to co-operate with the occupation authorities. Second, the Soviet forces used sheer terror tactics to quell rebellions against their rule. These included population transfers, assassinations and depopulation of the intensive areas of resistance in Central Asia. Third, they isolated the insurgents from outside support. Geopolitical factors and international power arrangements had on previous occasions worked in favour of the Soviet Union and successful insulation of the resistance in Central Asia, the Baltic and Eastern Europe had diminished the wars of resistance against the Soviet occupation. Encouraged by its successes elsewhere, the Soviet Union marched into Afghanistan quite confident of its power to install and stabilize a government of its own choice.

The Soviet military crossing into Afghanistan was quite in line with traditional Soviet operational strategy, focusing on the rapid movement of large formations of men and armour, as before with Budapest and Prague, but without thought for Afghanistan's peculiarities. The forward deployment and grouping of the various elements of the 40th Army and its support assets in Turkestan military districts, which began sometime in September 1979,[6] went largely unnoticed in Afghanistan. The arrival of the first formations of air assault troops and units from the 103rd Guards deployed from Bagram, and airlifts to Kabul between 22-26 December, surprised the Amin loyalists in the Afghan armed forces.[7] When they detected the chain landings of Soviet troops and their movement towards the strategic zones, it was already too late to act. There was some unwillingness on the part of the Afghan forces to give a determined fight—the Afghan army's 4th Armoured Brigade, even when ordered to protect the Amin regime, failed to move. The Soviet security personnel and advisers already stationed with the Afghan forces effectively pre-empted the possibility of a large-scale revolt by disarming suspected nationalist officers, and, in certain situations, sabotaging the mobility of the troops.[8] There were only a few instances of Afghan soldiers engaging the interventionist troops at Bala Hissar and Kargha.[9] By and large, Soviet forces accomplished the first task of seizing control of key military and political targets in and around Kabul without much resistance, then, after assassinating Amin and installing Karmal, they set out according to plan to capture strategic areas such as military installations, airports, bases and important urban centres. The bulk of the Soviet motorised divisions, meanwhile, crossed the Amu Darya from Termez in Central Asia.

The consolidation and extension of military control required a strong presence in all strategically vital areas of Afghanistan. From an operational standpoint, the country was divided into seven military regions. The security of Kabul and the surrounding areas was assigned to the 105th Airborne Division and the 360th Motorized Rifle Division. The north-eastern region was placed under the security of the 201st Motorized Rifle Division which was based at Kunduz and Faizabad. The 16th Motorized Rifle Division, located at Mazar-i-Sharif, had the primary responsibility for the security of Balkh province and surrounding areas. Herat and western Afghanistan were put under the protection of the 54th and 68th Divisions. The 275th Division, stationed in Jalalabad, was responsible for the east-central region. The 357th Motorized Rifle Division, operating out of Qandahar, covered the entire south-east.[10] The Soviets established two major operational headquarters, at Kunduz in the north and at Kabul's Taybeg palace. The logistics operations were run from Termez inside Soviet territory. Additionally, a transit base was set up at Kilagai.

COMPOSITION

Owing to the fierce resistance by the Mujahideen and the need to refine military tactics, the Soviets continuously introduced changes in the size, equipment and organizational structure of their forces. At any given time, with variation in the strength of various elements, the occupation forces had three Motorized Rifle Divisions, two independent Motor Rifle Brigades, two independent Motor Rifle Regiments, one Airborne Division, one independent Air-Assault Brigade, one independent Air-Assault Regiment and three *Spetznaz* Brigades.[11] A variety of other units included transport, construction, engineer, signal, artillery and airfield defence.

The occupation forces had two distinct categories, referred to as regular combat and support units. The larger part of the Soviet forces was designated to perform those non-combat tasks which are regarded as necessary to initiate or support military operations. Fighting a low-intensity protracted war against the Afghan guerrillas required maintenance and improvement of airfields, bases, barracks and roads for efficiency. That is why the underdeveloped infrastructure in Afghanistan consumed the greater part of the non-military assets of the Soviet forces. Much was invested in independent transport units, water-supply battalions, service and repair battalions, pipeline units, road-repair units, medical battalions, and construction troops. Extensive road-mining and rocket attacks by the Mujahideen against the valuable military targets not only made the task of support units hazardous but also kept them perpetually engaged in rebuilding.

The regular combat forces could be further divided into units earmarked for defence and security-related tasks and those designed for offensive counter-insurgency operations. The first category included all motorized rifle divisions except the reconnaissance subunits which were attached for special purposes. These forces were structured for the dual task of security of installations and defence of military outposts. Besides providing protection to civilian and military bases, airfields and logistic centres, they scouted the transport arteries. Mechanized infantry units regularly accompanied the large-scale convoys to ward off the frequent ambushes by the Mujahideen.

Another role assigned to the non-counter-insurgency combat units was the static defence of small military posts constructed along the major transit and supply lines. Many of these posts manned by infantry units were isolated and vulnerable to rebel attacks. In static mode, mechanized infantry was merely required to hold its ground and not go after the enemy beyond the security zones. Mechanized troops, however, did form a significant part of combined operations to deny territory, regain control of lost areas and battle for the security of besieged garrisons.

Effective territorial occupation of the whole country did not figure prominently in the Soviet military plans in Afghanistan. Such an undertaking would not only be costly and risky but would also require the deployment of much larger forces—perhaps three times the size of the occupation forces, which totalled approximately 105,000 to 115,000.[12] During the first year of war, the Soviets increasingly realized that a quick victory was not possible. Defections and the unreliability of the Afghan armed forces[13] also influenced Soviet plans for an all-out offensive strategy against the Mujahideen. Consequently, towards the end of the first year of war, the Soviets shifted their strategy from controlling cities and military installations to attacking Mujahideen bases. This change in objectives affected the composition of the main counter-insurgency forces.

A clear functional delineation among all the counter-insurgency forces in any operational theatre is not possible, much less in a war that has experienced the naked use of high-technology performance weapons. An assessment of Soviet military operations and weaponry shows that the Soviet military used three main types of counter-insurgency units with overlapping tasks. They were: a) airborne, b) air assault, and c) reconnaissance.[14] What makes these forces unique and distinguished is the strict criteria of selection, physical prowess and rigorous, specialized training. The heavy reliance on these forces in the Afghan war can be judged from the fact that while the counter-insurgency forces constituted only twenty per cent of the occupation forces, they received seventy-five per cent of the Hero of the Soviet Union awards.[15]

The airborne and air–assault units, known for efficiency, mobility and surprise in the conduct of operations, were widely used in rescue operations and attacks against the Mujahideen strongholds. They also mined the Mujahideen supply routes, set up ambushes on resistance groups and opened the way for Soviet convoys through passes. Frequent airborne and, depending on the nature of a mission, dismounted operations were conducted by these forces against villages and rural agriculture. Small-scale high-risk missions were assigned to the reconnaissance troops. They were regarded as the best among the counter-insurgency forces for their ability to undertake missions in the Mujahideen-held areas. A reconnaissance company was attached to every regiment and a reconnaissance battalion to each division. They were given the task of setting up fortified observation posts to watch some of the Mujahideen infiltration points.

It would be useful to offer some comments on the counter-insurgency equipment which the Soviet Union designed and used for the first time in Afghanistan. Within two years of their intervention the Soviets had introduced new weapons to enhance combat efficiency in the operational conditions of Afghanistan. Changes in equipment continued with the level of escalation, and in response to guerrilla strategies. In addition to the conventional hardware with which the Soviet forces moved in, the counter-insurgency troops acquired specialized equipment. A new AK-74 high velocity rifle of 5.45mm with well-designed bullets and image-intensifier was introduced. A new AKR-Krinko machine-gun was for the first time used in Afghanistan. A disposable grenade launcher (RPG-18) was another weapon tested in Afghanistan. Automatic 81mm mortar bombs for rapid and sustained fire against Mujahideen strongholds proved highly effective. The Soviet forces also used 152mm artillery shells with 'flechette' rounds containing razor-sharp slivers of steel.

The Soviet Air Arm for some time used the feared MI-24 (Hind) gunships in conjunction with SU-25 (Frogfoot) fighters. A new transport helicopter, MI-26, made its debut in the Afghan war. For mobility and rapid attack by mechanized infantry, the BTR-60 armoured personnel carrier was found very reliable. The BMP-2 armoured fighting vehicle, armed with 30mm cannon, which is an improved version of the BMP-1, was another innovation to suit the combat situation of Afghanistan. 'Butterfly' anti-personnel mines scattered in millions around the country continue to kill and maim Afghans. By no means is this an exhaustive list of the Soviet war inventories in Afghanistan; it simply suggests that the Soviets not only tested new weapons but also designed them to experiment with new operational concepts in their counter-insurgency war.

OPERATIONS AND TACTICS

As the war progressed and the Soviets learned their lessons, both the overall strategy and military operations underwent a change. The counter-insurgency doctrine and equipment were modified to adapt to the geographic and topographical conditions of Afghanistan. The Soviet planners increasingly realized the importance of airborne assaults and close air support which roughly substituted for the artillery supporting fire in mountain warfare. The Soviet counter-insurgency performance undoubtedly improved with the concentration on air-mobile operations. Total dominance in the air and the effectiveness of helicopter-gunships gave a key advantage to the Soviet forces. Within a short time the helicopter became the mainstay of Soviet combat. It was used in multiple roles: a) transporting specially-designed armoured fighting vehicles and self-propelled guns for mechanized infantry close to the operational theatres; b) surprise air assaults against the Mujahideen positions; c) shifting of troops and d) in situations when the guerrillas were found exposed in flatlands, as a platform for sniping. The rapid movement, manoeuvrability and firepower of the MI-24 and MI-8 inflicted heavy damage on the resistance and highly demoralized its fighters. It was not until the acquisition of Blowpipe and Stinger missiles by the Mujahideen that the Soviet helicopter-borne operations and aerial bombardment became costly and ineffective.[16]

Constrained by logistics and the combat environment, the Soviet strategy avoided both over-extending its military presence in the vast countryside and attaching prestige to securing territories. Rather, the Soviet military planners combined flexibility with aggressive posturing in areas that presented a major threat to strategic points and communications. As the Soviets located their garrisons close to major bases for rapid response and reinforcements, they launched aggressive operations to dislodge the guerrillas from surrounding areas with the aim of keeping them at a less threatening distance.

The Soviets, as any wise counter-insurgency force would be, were highly sensitive to casualties. They sent out forces only under favourable conditions. Most of the punitive and dislodging operations were undertaken only when the Mujahideen moved close to communication lines or posed a threat to valuable targets. In most of the anti-guerrilla sweeps, concern for the safety of Soviet lives was obvious. Most of the assault operations were planned in such a manner that the Afghan army units and militia would be deployed in forward positions. In the first phase of these operations, the Soviets would bombard the area by aircraft and use heavy artillery for softening up the target, after which the troops would march in for combing. At times, when the reliability of the Afghan soldiers was doubted, the Soviets moved with them in close formation. But invariably, Soviet landings close to operations or forward movements followed the indiscriminate use of long-range weapons. The assaulting infantry used light tanks and armoured vehicles for fast and relatively secure attacks.

The Soviets amended some of their previous operational doctrines by emphasizing decentralized command and control and small-scale offensives.[17] These modifications were influenced by detached operations in the mountains where it was almost impossible to adhere strictly to plan or to constantly seek advice in operational contingencies. Due to poor intelligence, the Soviet assault forces found themselves in situations quite different from those visualized by the senior officers at headquarters. Although the overall counter-insurgency doctrine remained dominant, the Soviets changed operational concepts more by adaptation than by introducing any revolutionary change. Most of the doctrines and concepts tried in Vietnam and elsewhere were learned by the Soviet forces and applied in Afghanistan. Despite the use of new technology and a different military organization, the Soviet counter-insurgency experience hardly broke any new ground. The classical learning in this type of warfare would continue to guide military minds.

The Soviet operations, though varying in size, objectives and the nature of targets, might be roughly categorized into four groups. There was hardly any time-sequence or consistency in strategic planning. Shaped by the initiatives of the Afghan guerrillas, the Soviet operations were largely reactive. Faced with reality, quick victory slid down the ladder of Soviet objectives. But this does not mean that they did not try hard to win the war. The 'limited' forces and 'limited' war presented an all-out war scenario for the Afghans. Various phases in the escalation of conflict and the use of overkill power made the war extremely costly in human and material terms for the Afghan nation.

a) Dislodging Campaigns

After basing securely in the garrisons around the country, the Soviet forces began launching attacks against the areas held by the Mujahideen. Major cities, such as Herat and Qandahar, or strategic points, whenever occupied by the Mujahideen, were repeatedly attacked by the Soviet forces.[18] Ejection of the guerrillas from major towns was much easier, because of the collateral damage to the civilian population, than from the valleys.

The Soviet offensives in the Panjsher valley were primarily aimed at vacating an area long held by guerrillas under the command of Ahmad Shah Masud. The Soviet forces struck Panjsher no less than eleven times, each time using a different combination of weapons and tactics. Masud and his forces, despite suffering considerable damage during each invasion, survived primarily because they avoided set-piece conventional battles. Rather, they retreated safely to the side valleys and mountain tops, or on occasion vacated the valley altogether. After each attack, the Soviet force declared a victory and left the main bed of the valley, confining itself to patrols mainly by helicopter-landed forces. However, in 1985, when the Soviet forces left the Afghan troops to maintain a presence in Pechgur, they were ambushed by Masud's men. It was one of Masud's major victories. The Central Corps Chief of Staff, Brigadier Ahmaduddin, and one of the six colonels present in the base were killed.[19] The entire post was overrun following the surrender of 110 officers and 350 conscripts.

The Panjsher valley experienced repeated dislodging assaults primarily for two reasons. First, the valley and the Mujahideen presence was close to the Salang tunnel and the highway from the Soviet border to Kabul, and the Soviet supply convoys from Central Asia were repeatedly ambushed by guerrillas from this valley.[20] Second, the concentration of one of the best-trained Mujahideen forces under the command of Masud, probably the ablest fighter in Afghanistan to date, was a constant threat to the Kabul regime. Using the valley as a fortress, Masud attracted area commanders and fighters from all over the country. Training of the guerrilla forces, expansion of communications networks among the area commanders and better co-ordination of insurgency, particularly in the northern provinces, provoked a variety of punitive and dislodging campaigns by the Soviet forces. International media publicity about the Panjsher battles added to the prestige which the Soviet forces had attached to the occupation of the valley.

The Soviet dislodging operations in the Panjsher valley and elsewhere followed a common pattern—with variations in the scale of the offensives, the troops moved first to cordon off escape routes. Blocking groups of paratroops were inserted by helicopter in passes where mechanized infantry had difficulty in moving its armour.[21] Aerial reconnaissance for locating targets preceded air strikes by SU-25s, SU-17s and MIG-21s which dropped high-explosive bombs. For low-level raids retard bombs were used. The dropping of RBK-250 cluster bombs was reported in Kunar province. The bombing of the areas of operations went on even for weeks before helicopter-borne landings started or ground forces advanced. Within a large operational zone such as Panjsher, mini-offensives were conducted after gaining control of safer ground and the consolidation of forces.

b) Depopulation

In the guerrilla-controlled areas, the Soviets unleashed destruction second only to the US strategic bombing in Vietnam. Intensive bombing of the rural areas, targeting the villages, and extensive ground search-and-destroy missions, produced one of the largest refugee outflows of the post-World War II era.[23] Within only a few years over three million Afghans from all parts of the country swarmed the borderlands of Pakistan, over one million poured into Iran and another two million became internal refugees seeking food and security in the major towns.[24] The eastern provinces adjoining Pakistan suffered the heaviest loss of population. It was the popular support for the Mujahideen groups by the Pushtun tribes that invited the wrath of the Soviet forces. This is not to suggest that civilian targets in other provinces were spared. In August 1982, the Panjsher valley witnessed one of the extensive campaigns against villages which triggered a major refugee flow.[25]

The shooting of civilians and the deliberate razing of villages, which are well-established in the literature[26] and need no repetition here, were partly a manifestation of indiscipline and frustration.[27] Marauding and the excessive use of violence in many counter-insurgency wars occur largely out of a sense of superiority which the soldiers of organized armies usually display. In the case of Afghanistan, one may not rule out a centrally-organized strategy aimed at what Louis Dupree described as 'migratory genocide.' The transfer of population was a part of the Soviet counter-insurgency doctrine which had been practiced before in its wars in Central Asia. An effective counter-insurgency war could not be fought against guerrillas swimming in a sea of a supportive population. The Kabul regime, having failed to win the hearts and minds of the Afghans, collaborated with the Soviet forces in depopulating areas of tough resistance.[23] Obviously such a policy was aimed at: a) separating the guerrillas from the population, thus making them more vulnerable, b) eliminating the intelligence support from the villagers, and c) denying them sources of food.

The destabilization of Afghanistan's traditional demographic structure has not only produced one of the most tragic events in contemporary history but also undoubtedly presents a costly and difficult task for any future reconstruction in the country.

c) Destruction of Rural Infrastructure

Along with the horrors of war, what threatened the civilian population was the prospect of famine. To expedite the migration of the population, the Soviet forces cut vegetation, burnt crops and set the forests on fire. The irrigation system, mostly natural though in some areas man-made, held a delicate balance in the rural economy. Not only by extensive indiscriminate bombings but also by well-calculated acts, the Soviet forces targeted small dams, water channels and mountain streams. Many of these 'scorched earth' operations accompanied attacks on villages.

Afghanistan's ecological balance has been gravely disturbed. No amount of foreign aid will fully restore its pre-war environmental setting. Very little exists today in terms of regenerating agricultural productivity to support the same level of population. One reason refugees in Iran and Pakistan are reluctant to go back is that they foresee nothing to support their families. Perhaps sustained and determined efforts with the involvement of international agencies might help restructure the means of production in largely rural Afghanistan.

Much like human life, animal life in Afghanistan has been devastated. Millions of animals were either abandoned in haste or taken to Iran and Pakistan by the refugee families. Since the refugees faced competition with and hostility from the local inhabitants over grazing lands, and since there was an economic need, they sold their stock. Many of the refugees, particularly in Pakistan, took up new jobs in construction, transport and factories. Shepherding might still be popular among the Afghans but lack of infrastructural support discourages them from reinvesting in this old profession. Another factor that took a heavy toll of animal life was the carpet bombing of valleys[29] and the dropping of butterfly mines (PFM-1) along all the passes frequently crossed by the Mujahideen caravans. Mining of areas around highways, garrisons and military facilities and scattering of anti-personnel mines in the fields maimed tens of thousands of men, women and children and killed a large number of animals.

One of the most tragic casualties of war has been the rural health centres and village schools, which even according to the standards of developing countries were inadequate. The Marxist regime and the Mujahideen leadership have each accused the other of willfully destroying buildings meant for treating patients and imparting knowledge to children. No matter what the truth of these counter-allegations, the fact is that the Soviet war has caused tremendous destruction of assets in the rural areas which resource-strapped Afghanistan will find costly to rebuild.

d) Interdiction and Blocking Operations

The disruption of the guerrillas' infrastructure and supply lines in areas bordering Pakistan never receded in the strategic priorities of the Soviet Union. The Soviet and Afghan forces began making sweeps as soon as they had tightened control over the points vital to the mobility and shifting around of troops.[30] A major change in Soviet war strategy occurred in 1983 with the mobilization of a large security force in the border provinces. Hundreds of tanks and combat vehicles were used in conducting operations.[31] The interdiction on guerrillas crossing over from Pakistan with better weapons became central to the Kabul-Moscow counter-insurgency efforts. No matter how masterfully planned, the blocking operations could not succeed against a historically porous border, difficult terrain and logistical problems. Quick forays into Mujahideen-held areas and temporary evictions hardly stopped the movement of the guerrillas or their supplies. Soviet assaults constrained the guerrillas only for the limited time that a strong deployment of security forces could be maintained.

In 1985, frustrated by the regular infiltration of guerrillas from their sanctuaries in Pakistan, the Soviets advocated more aggressive campaigns along the borders, and their determination to break guerrilla power was apparent from the large-scale involvement of their forces. The operations in Kunar[32], including the capture of Barikot in 1985, had three objectives: a) eviction of the enemy from areas which they had used as a safe base for menacing the Soviet garrisons and road convoys, b) establishment of a military presence close to guerrilla supply lines for quicker response, c) permanent damage to the guerrilla infrastructure in the borderlands. Intensified military actions near the Pakistan border continued till the end of 1987.[33]

The Soviets applied a combination of tactics to 'seal' the borders, but none could produce lasting effects. Buying of the tribes with incentives of money, weapons and security was tried from the outset of the war.[34] The Soviet occupation politics, to be discussed fully in Chapter 6, emphasized the denial of sanctuaries to the Mujahideen in the tribal areas both inside Afghanistan and along the frontier region of Pakistan. The Kabul-Soviet successes were limited as the tribes played a dual role, wresting concessions from both sides. Some reports suggested that special units generally known as *spetznaz* operated from the outposts, dressing as Mujahideen and adopting gang-tactics. They terrorized the villages to generate hostility against the guerrillas.[35] Also, by attacking rival Mujahideen groups, they exacerbated tensions and disunity among the guerrillas.

Staging ambushes on the guerrilla bands along the supply trails was another common practice of Soviet forces stationed in the eastern provinces. The air-assault forces dropped in the upper reaches of the high passes operated quite effectively using long-range assault weapons. But such temporary ambush stations were few and isolated and thus vulnerable to the laying of sieges by the Mujahideen. One of the most dangerous and effective tactics the Soviets used against infiltrating guerrillas was the air-dropping of anti-personnel mines by helicopters along the guerrilla routes. Millions of mines widely scattered in the region did make the night-time movement of guerrillas slow, costly and risky.

e) Use of Chemical Weapons

The alleged use of chemical weapons by the Soviet Union in Afghanistan has been a controversial and widely debated subject with no firm conclusions. In an atmosphere of political acrimony, the alleged use of chemical agents could neither be confirmed according to strict scientific criteria nor rejected outright as a propaganda ploy. A number of factors reinforce the suspicion that Soviet forces might have resorted to chemical warfare in some of their counter-insurgency campaigns. First, the Soviet chemical warfare capabilities and Soviet military doctrine which saw chemical agents as an integral part of its defence policy. In Soviet military parlance, chemical weapons were better suited for surprise and for seizing a specific military target. Western sources have estimated the strength of the Soviet Union's chemical troops in the range of 80,000 to 100,000, with the largest stock of chemical munitions in the world.[36] A variety of agents such as mustard gas, tabun, soman, and tricothecene mycotoxins were developed and deployed in the payloads of Soviet conventional munitions. The offensive chemical operations in Soviet military exercises by conventional front-line units were widely feared.

Second, the Soviet forces that invaded Afghanistan had a full complement of their normal Warsaw Pact contingency equipment. The Soviet formation in Afghanistan included decontamination vehicles and AGU-3 detoxification chambers for decontaminating personnel. The presence of some of these units in forward combat areas further strengthens the reports.[37] Defectors from the Afghan army gave details of the units designated for chemical warfare.[38] A number of accounts reveal the existence of the 234th Gas and Chemical Department headquartered at Qal-a-Janki close to Soviet residential quarters in Kabul.[39] Supplies of chemical warheads for artillery shells, helicopter-borne cannisters, and gas-filled infantry grenades were reported. The formation of chemical field battalions and the training of Afghan army officers in the Soviet Union for chemical warfare was widely narrated by defectors.[40]

Third, the reports of chemical injuries and mysterious deaths were consistent, mostly from areas of strong Mujahideen presence. Afghan fighters and refugees repeatedly told of yellow dust and coloured smoke produced by bombs.

The survivors of attacks and witnesses who escaped experienced painful burning of the eyes and throat and vomiting of blood. They reported skin falling off corpses after turning black. In some cases the toxins caused suffocation and incapacitation of victims.[41] Some organizations sympathetic to the Afghan resistance drew world public attention by widely publicizing reports of chemical warfare. Also, the US State Department issued a strong report concerning the use of mycotoxins by the Afghan-Soviet forces.[42] The Soviets initially attempted to block the formation of a UN Commission to investigate the situation in Southeast Asia and Afghanistan, then when it was formed, refused to co-operate. Moscow's obstructionist attitude strengthened the doubts further. A relatively convincing document on the subject is a study prepared by the Australian jurist Felix Ermacora for the UN[43] which accuses the Soviet Afghan forces of having used chemical weapons against the resistance. A number of independent inquiries corroborated the evidence collected in the UN report.

Many observers have questioned the eyewitness reports and circumstantial evidence supporting the alleged use of chemical agents. They have tended to see the allegations as malign and propaganda-oriented.[44] The issue remains unresolved for two reasons. First, no hard physical evidence has been retrieved beyond the gas masks found on Soviet corpses. Second, the sketchy details given by refugees unfamiliar with complex substances should be treated with care and a great measure of doubt. In view of the inconclusive evidence, but in the face of reports that have persisted for quite a few years, it becomes difficult to argue in favour of either side with certainty. In the end, Moscow has escaped less than unscathed by the allegations despite being given the benefit of the doubt. The Mujahideen sources have vehemently argued that there was no reason for the Soviet forces to observe restraint in the use of chemical agents when they proceeded to inflict about two million casualties on the Afghans, and showed no scruples in scattering millions of anti-personnel mines, which was no less a violation of international law than the use of chemical weapons was of the Geneva Protocol. Whatever the truth of the matter, the issue touched sensitive chords in international public opinion.

LIMITATIONS AND PERFORMANCE

When the Soviet forces intervened in Afghanistan, neither Moscow nor the West expected that military operations against the ill-equipped and unorganized resistance would be required beyond six months.[45] In the event, the Soviet forces got locked into a protracted war which the planners had not foreseen. The objective conditions in Afghanistan were ignored by the occupation forces—which took a heavy toll of their men, material and image. It was never realized by Moscow that the crisis in Afghanistan was primarily a political struggle between indigenous antagonists, with one group fighting for political survival and the other opposing its programme and connections with Moscow. From the outset, Soviet ideology and its physical presence were counter-productive in resolving the crisis in favour of the Afghan Marxists. The presence of the Soviet forces and their campaigns against the local population were provocative.[46] As a result, the country plunged into total chaos and the urge to rebel further strengthened the forces opposed to the Marxist regime.

A major lesson which many resistance wars have taught is that the military strength of the conquerors and their technology are not always the final arbiter. The guile of the opponents, the level of national mobilization among the subject peoples, their commitment to freedom and willingness to endure extreme hardship and death, along with the physical conditions, can offset the material advantages of interventionist forces. Vietnam and Afghanistan showed that physical presence neither ensured political consolidation nor promoted the resolution of internal conflicts.

Another issue relevant to an examination of the limitations and performance of the Soviet forces in Afghanistan is domestic and international pressure. In modern times, the military must address political considerations when tempted to break battlefield stalemates by escalating combat to an unacceptable level. The Soviet violence in Afghanistan did reach a high level, but markedly less than what Moscow was capable of inflicting. One must dispel the notion that owing to the closed nature of the political order and with the image of ruthlessness, a power such as the Soviet Union could continue a war indefinitely.[47] Finally, it had to count the dead, have a second look at the cost-benefit equation and mend its image in the Islamic countries and the Third World.

The performance of the Soviet Army in Afghanistan must be judged against the objectives it had to achieve. The obscurity of the objectives at the different levels of escalation adversely affected both the planning of counter-insurgency strategy as well as the battle performance of the forces. The Soviet army attempted to pursue a range of goals within the broad objective of pacifying the country, which has raised a number of questions. Was the political climate in the country suitable for the efficient military role of a foreign power? Were the structure and strength of the forces adequate to crush Afghan resistance? Could a modern army, organized, trained and equipped to fight high-intensity wars, be prepared for a ferocious popular resistance? Although the Soviets continued to claim that they were not beaten in Afghanistan and were not driven out by force, yet the fact remains that the war was stalemated with no victory in sight. Partly it was frustration with the war and the no-win situation that made Moscow call it a day. Afghanistan was a wound that had been bleeding the Soviet Union for nearly nine years.[48]

Moscow later admitted that intervention in Afghanistan was a mistake.[49] It has disclosed, though belatedly, that it suffered more than 13,000 dead and 330 still unaccounted for, some of whom are believed to be held by the Mujahideen.[50] Western and even some Soviet sources have cited the Soviet casualty rates as being much higher than was officially admitted in 1989.[51] No less was the damage to aircraft, armoured vehicles and other war materials.[52] The economic cost of the war was put by former Premier Nikolai Ryzhkov at five billion roubles a year (US $758 million according to the official exchange rates of 1989).[53] Saner elements in the Soviet establishment must have paused and posed themselves the question: was the Afghan adventure worth the cost and sufferings the Soviet citizens were made to bear? It was partly the negative response to this question that prompted the Kremlin's reappraisal of the reality in Afghanistan. Undoubtedly, the Soviet disengagement from Afghanistan was as much the result of 'new thinking' in the Soviet Union[54] as of the stalemate in the conflict.

Having noted some conceptual and structural constraints on the Soviet military performance, it would be equally important to look at a number of problems that the Soviet army encountered. First, the Soviets found the Afghan army, which they had equipped and trained for twenty-five years, incapable of launching offensive operations. In the early phase of the Soviet intervention, Afghan troops defected to the resistance, with their weapons, in large numbers. The Afghan army's strength fell by September 1980 to just 30,000.[55] Defections from the Afghan army significantly reduced its participation in the war for the first few years. In the beginning, every Soviet battalion engaged in operations included at least two Afghan units of roughly similar size. In an attempt to raise the army to its optimal size, the Kabul regime reduced the training period for commissioned officers from five years to just three months. The quality and reliability of the Afghan forces constantly worried the Soviet officers and soldiers. They feared sabotage, leaking of operational plans and even ambushes. On occasion, the Afghan troops let the Mujahideen escape when bottled up.[56] A large number of Soviet aircraft were sabotaged by Afghan soldiers. The Soviets gradually rebuilt the Afghan army and left it in better shape than they had found it at the start of the war, but the morale of the Afghan army remained low, and it continued to be heavily infiltrated by the Mujahideen groups.[57]

Second, in the initial stage, the Soviet units comprised a large number of local reservists from Central Asia who had neither enough military skills nor the motivation to kill fellow Afghan Muslims.[58] Although the original Central-Asian Muslim units fighting in Afghanistan were withdrawn after a few months, they continued to form a significant part of the occupation forces.[59] Even the regular Soviet forces had limited training for a counter-insurgency war in desert and mountain conditions. Only the special-purpose units had been trained for Afghanistan-like situations. The Soviet strategists were quick to realize the training inadequacy, and set up mountain-warfare training centers in Azerbaijan and Tajikistan; overall, however, training remained poor and insufficient for the topographical conditions of Afghanistan.

Third, the Soviet soldiers had morale and motivation problems. The crude and uninspired official propaganda widened the perceptual gap between what they were told about Afghanistan and what they saw at their first contact. It did not take them much time to find that they were fighting against poverty-stricken Afghans, not 'mercenaries' from China and the United States. In situations of aimless killing, it is usual for soldiers to develop psychological disorders. Alienation, use of drugs, violence and crimes including brutal murders of fellow-countrymen have been well documented.[60] Substantial evidence to this effect came in November 1989, when the Soviet Legislature granted an unconditional amnesty to 2,540 former servicemen convicted of crimes, including 200 serving terms for premeditated murder.[61] The aggravated circumstances in which the crimes were committed and the pressure of wartime situations were cited as the basis for showing mercy to the convicts.

Finally, the Soviet forces faced problems, as would any other occupation force, in unfamiliar combat conditions. Although some very intensive groundwork had been done by the Soviet military planners for fighting a war in Afghanistan, it took them a long time to gain sufficient knowledge of the operational environment.[62] Quite often mechanical failures of equipment occurred, and maintenance facilities were inadequate. Climatic conditions, poor organization and the generally low quality of leadership compounded the problems. Corruption in the military establishment,[63] particularly in distant detachments, and rivalries between old and new soldiers,[64] and verbal and physical abuse of subordinates demoralized the troops and considerably undermined their ability to fight a physical war of attrition in a foreign land. Poor food, illness[65] and drugs further reduced the combat effectiveness of the forces.

The Soviets lost much time in recognizing some of the shortcomings and taking remedial measures. As indicated earlier, special training centres were established for mountain and desert warfare. Better and new equipment more suited for counter-insurgency operations was acquired. Special anti-sniper squads were formed. Heavy investment was made in improving logistics, communications and other military infrastructure. For all practical purposes, the operations in Afghanistan were integrated with the Turkestan military district. Not only did the air force locate its aircraft there on a rotational basis,[66] but major repair and maintenance facilities were also built to service the war-machine in Afghanistan. Quite often special troops were flown directly from Central Asia for conducting operations in northern Afghanistan. The high-level carpet bombing of the Panjsher valley was done by aircraft operating from the Turkestan military district.[67]

Gradually, the Soviets transformed their Afghanistan army from a high-intensity mechanized force to one suited to the requirements of anti-guerrilla warfare. Although no significant changes were made in the command and organizational structure, which is known for its over-centralized planning and decision-making, greater tactical flexibility was achieved with the restructuring of the combat elements. The motor rifle divisions were reduced and reformed. The advantages of air mobility and helicopter gunships were realized in the early stages of war. This is shown by the massive rise in the number of Soviet helicopters from 60 in June 1980 to around 300 in 1981, which operated from Bagram, Kunduz, and Qandahar.[68] The number of jet fighters jumped to 130, comprising mainly Mig-21s, Mig-23s, and SU-17s. Air assault and helicopter-gunships proved more effective than any other anti-guerrilla tactics.

The changes in operational strategy were neither sufficient nor timely enough to regain the initiative. Time is always a critical factor in the outcome of insurgencies. If not insulated quickly, wars of resistance take a heavy toll of the occupation forces. The resilience of the Afghan resistance and the no-win situation for the Soviet Union allowed considerable opportunities to the rival superpower, the US, and to neighbouring Pakistan to increase the scale of war. The failure of the Soviet Union to efficiently prosecute its counter-insurgency in Afghanistan is rooted in its failure to isolate the Afghan resistance. The Mujahideen had a viable basis of external support and they demonstrated that they could carry on their struggle against the Kabul-Soviet forces indefinitely.

The Soviet Army in Afghanistan achieved one remarkable success. Although it could not destroy the resistance, it effectively prevented the anti-regime forces from capturing and holding any strategic target. The political and military arrangements which the Soviet Union had helped to build during its ten years of occupation withstood its withdrawal from Afghanistan for three years. But we must note the fact that neither Soviet interest nor its material support to the Kabul regime diminished[69] until the disintegration of the Soviet Union itself in December 1991. Enormous Soviet military supplies to the Kabul regime continued on a regular basis, denying the Mujahideen any significant military success.[70] Also, the Soviet protective umbrella, held over the regime for some nine years, gave it the ability to reorganize the army and launch more aggressive political programmes to broaden the support base. Moreover, they left the Afghan armed forces better equipped and more disciplined than they had found them at the time of invasion.

It was generally thought that, following the Soviet troop pull-out in February 1989, the Marxist regime would not be able to defend itself. Most of the analysts of the Afghan scene who held this view tended to discount the Soviets' indirect support for the regime, which was no less critical in maintaining its power than the direct military involvement. Thousands of Soviet advisers were left behind to command and conduct the counter-insurgency campaigns. Hundreds of millions of roubles-worth of military hardware kept coming to the regime each month, besides adequate supplies of food and fuels. It was the failure of the *coup* by the communist hard-liners in Moscow in August 1991 that began to expose the vulnerability of the Kabul regime and the degree of its dependence on its Soviet patrons. Once supplies of arms to Kabul stopped from 1 January 1991, and the economic crisis within the former Soviet Union itself made it practically impossible to meet previous commitments of food and fuels to Kabul, the regime began to crumble from within. It is remarkable that, lacking Moscow's support, the Marxist regime headed by Najibullah did not survive beyond 105 days, and it finally collapsed on 15 April 1992.[71]

NOTES

1 The Soviet forces had SAM-7B anti-aircraft missiles and ZSU-23 automatic anti-aircraft cannons.

2 Alexander Prokhanov, 'Why the Soviets failed in Afghanistan,' *Globe* (Karachi), June-July 1989, pp. 22-4.

3 Lawrence Lifschultz, 'Voices from Moscow,' *Herald* (Karachi), July 1987, p. 29.

4 For a comparative analysis, *See,* Michael G Fry and Condoleezza Rice, 'The Hungarian Crisis of 1956: The Soviet Decision,' *Studies in Comparative Communism,* Vol. 16, nos. 1-2, Spring/Summer 1983, pp. 85-98; Fred G Eidlin, *The Logic of 'Normalization': The Soviet Intervention in Czechoslovakia of 21 August 1968 and the Czechoslovakian Response* (Boulder, Colo.: East European Monographs, 1980); Jiri Valenta, *Soviet Intervention in Czechoslovakia* (Baltimore, Md.: Johns Hopkins University Press, 1978); Ronald R Pope, 'Miscalculation in Soviet Foreign Policy (With Special Reference to Afghanistan),' *Crossroads* No. 8, Winter/Spring 1982, p. 107-24.

5 Stephen Blank, 'Soviet Low-Intensity Conflicts: An Agenda for Research,' *Conflict,* Vol. 9, No. 1, 1989, pp. 1-19.

6 On 20 September 1979, in Washington, the US State Department spokesman reported 'increased activity among the Soviet units north of the Afghan-Soviet border.' *International Herald Tribune,* 21 September 1979.

7 *Sunday Times,* 6 January 1980.

8 At Rishkoor and Kargha Soviet advisers locked up officers loyal to Amin and immobilized vehicles. *Los Angeles Times,* 10 January 1980.

9 *International Herald Tribune,* 14 January 1980.

10 For this information I have benefitted from Zalmay Khalilzad, 'Moscow's Afghan War.' *Problems of Communism,* Vol. 35, January-February 1986, pp. 1-20.

11 For detail, *see,* Mark Urban, *War in Afghanistan* (New York: St. Martin's Press, 1988), pp. 229-32.

12 *The Economist,* 26 October 1985, p. 47.

13 In the early phase of Soviet intervention a significant portion of the Afghan forces defected with heavy weapons. *Asia Week,* 30 May 1980.

14 For detail of these categories and specific tasks, *see,* David C Isby, 'The Vertical Threat: Air Assault and Air Mobile Brigades in the Soviet Army,' August 1985; Department of the Army, *The Soviet Army: Troops, Organization and Equipment,* (Washington, D.C., July 1984), pp. 31-2.

15 Alex Aleyiev, *Inside the Soviet Army in Afghanistan,* (Santa Monica: Rand Corporation, 1988), p. 26.

16 The hit ratio of the Stinger was 30-40 per cent. According to one estimate, by October 1986 270 Soviet aircraft had been destroyed by missiles. Aaron Karp, 'Blowpipes and Stingers in Afghanistan: One Year Later,' *Armed Forces Journal International,* September 1987.

17 Colonel Y Pavlov, 'The Commander and Modern Warfare, The Mountain Variant,' *Kraznaya Zvezda,* 1 June 1980.

18 *See,* a report by Tyler Marshall on Herat, 'Afghan City is Dying in Constant War,' *International Herald Tribune,* 5 September 1980.

19 Radio Kabul report, 28 June 1985.

20 Edward Girardet, 'Afghanistan's Che Repels Soviet Attacks,' *Guardian,* 11 November 1981.

21 *See,* a report on Soviet attacks on the Panjsher Valley in *Far Eastern Economic Review,* 2 July 1982.

22 David Isby, 'Soviet Tactics in the War in Afghanistan,' *Jane's Defence Review,* Vol. 4, No. 7, 1983.

23 On this subject *vide* two studies by Marek Sliwinski, 'Afghanistan: The Decimation of a People,' Orbis, Vol. 33, No. 1, Winter 1989, pp. 39-55 and 'On the Routes of 'Hijrat',' *Central Asian Survey,* Vol. 8, No. 4, 1989, pp. 63-93.

24 *Far Eastern Economic Review,* 1 February 1990, p. 30.

25 Damage in the valley was so extensive that the population dwindled from 80,000 to 50,000. Interview with the Jamiat representative in Islamabad on 27 December 1989.

26 *See, Tears, Blood and Cries: Human Rights in Afghanistan Since the Invasion 1979-84* (New York: Helsinki Watch, 1984). A number of Soviet prisoners have talked of indiscriminate slaughter of civilians during operations. 'Private Igor Rykov, Soviet Prisoner Interviewed,' Radio Free Europe-Radio Liberty, *Background Report* 220/84; 1 June 1984.

27 The court-martialling of a Soviet air-assault commander and JCOs on account of a revenge attack on a village was reported by a Soviet prisoner of the Mujahideen. 'Private G Dzamalbekov, Soviet Prisoner Interviewed,' RFE/RL, *Background Report,* RL/84, 12 July 1984.

28 Radek Sikorski, 'Frontline Afghanistan,' *Statesman* (New Delhi), 6 October 1986.

29 Edward Girardet reported that eighty per cent of the buildings in the Panjsher had been damaged or destroyed since the rumbling-up of the first Soviet vehicles there. See his 'Occupation Challenges Tradition,' *Guardian,* 10 November 1981; *International Herald Tribune,* 18 April 1984.

30 Kabul Radio report, 7 December 1981.

31 See reports in *Parade Magazine,* 24 February 1983, pp. 12-19; *Christian Science Monitor,* 30 March 1983; *Wall Street Journal,* 26 April 1983.

32 *Afghanistan Information Center Bulletin* (AIC) January 1985; *International Herald Tribune,* 4 June 1985.

33 *The Muslim,* 18 March 1987, 13 November 1987.

34 *AIC Bulletin,* No. 7, November 1981.

35 Commander Daulat interviewed in *AIC Bulletin,* February 1984.

36 On Soviet chemical warfare capabilities, training and doctrine, see E M Kallis, 'Chemical Warfare: Background and Issues,' Congressional Research Service, June 1981; John Erickson, 'The Soviet Union's Growing Arsenal of Chemical Warfare,' *Strategic Review,* Fall 1979.

37 James A Phillips, 'Moscow's Poison War: Mounting Evidence of Battlefield Atrocities,' *Backgrounder* (Washington, D.C.: The Heritage Foundation, 5 February 1982), p. 10.
38 *Daily Telegraph* (London), 29 October 1981.
39 John Fullerton, 'Poisoned Earth Policy,' *Far Eastern Economic Review*, 30 October 1981.
40 Author's interviews with defecting Afghan soldiers in Peshawar, October 1987.
41 *See*, reports attributed to resistance sources. *Statesman*, 8 April 1981: *Times of India*, 15 July 1981; *Sunday Statesman* (New Delhi), 27 September 1981.
42 US State Department, 'Reports of the Use of Chemical Weapons in Afghanistan, Laos and Kampuchea,' Summer 1980.
43 *The Muslim*, 14 November 1986. A report of the group of UN experts who had conducted interviews in Pakistan and collected evidence in 1982 confirmed the use of chemical weapons. *See*, *Wall Street Journal*, 7 June 1982.
44 *See*, Mark Urban, op. cit.
45 *See*, an assessment by Drew Middleton: 'Russian Troops in Afghanistan: Uncertain Road,' *New York Times*, 16 January 1980.
46 Selig Harrison reported 'widespread popular distaste for the Soviet presence.' *See*, his 'Moscow's Slow Hand in Kabul,' *Times of India*, 27 May 1984.
47 For the first six years, the Soviet coverage of war was limited and sang praises of heroic deeds. *New York Times*, 3 December 1984. But when a shift in policy under Gorbachev occurred and a relatively realistic picture of Afghanistan was portrayed in the media, opposition to the war mounted. *Nation* (Lahore), 11 July 1987.
48 Mikhail Gorbachev once termed Afghanistan a 'bleeding wound.' *The Economist*, 21 May 1988, p. 25.
49 *Dawn* (Karachi), 24 January 1983.
50 *See*, an interview with Colonel-General Grigorii Krivosheev, Deputy Chief of the General Staff of the Soviet Armed Forces, in *Kraznaya Zvezda*, 22 November 1989. Figures also published in *New York Times*, 29 November 1989, p. A10.
51 *The Muslim*, 15 January 1986, quoted a Western source putting the Soviet casualties at 30,000. An Estonian Komsomol newspaper, *Noorte Haal*, 24 January 1989, suggested that as many as 50,000 troops had died. *Radio Liberty Report*, RL157/89.
52 *See*, note 14.
53 *Far Eastern Economic Review*, 13 July 1989, p. 17.
54 The author's interview with Professor Yuri Gankovsky on 22 December 1989, in Islamabad, Pakistan.
55 US State Department, *The Soviet Dilemma in Afghanistan* (Washington, D.C.: June 1981).
56 Author's interviews with Afghan resistance leaders in Peshawar, 5 January 1990.
57 There have been numerous mutinies and *coup* attempts from within the army. One of the major attempts was crushed in December 1989. *Kabul Times*,

6 December 1989. Another one was attempted by the then defence minister, Shahnawaz Tanai, in March 1990.

58 Alexander R Alexiev, and S Enders Wimbush (eds.), *Ethnic Ministries in the Red Army: Asset or Liability* (Boulder, Colorado: Westview Press, 1988), part 5.

59 Alexander Alexiev, *Inside the Soviet Army in Afghanistan,* p. 10.

60 See details in ibid.

61 *New York Times,* 29 November 1989, p. A10.

62 Sgt. Yuri Grigorevich Povarnitsin and Pvt. Mohammed Yazkuliev Kuli interviewed by Alain Fandeux, 'Soviet Captives Describe Afghan War Roles, *Washington Post,* 28 September 1981.

63 See a report in *The Economist Foreign Report,* 30 July 1987.

64 Pvt. V Naumov interviewed in RFE/RL, *Background Paper,* RL/121/84, 19 March 1984.

65 *Sunday Telegraph,* 16 June 1985.

66 David Isby, 'Soviet Tactics in the War in Afghanistan,' *Janes' Defence Review,* Vol. 4, No. 7, 1983.

67 *International Herald Tribune,* 18 April 1984.

68 David Isby, 'Soviet Tactics in the War in Afghanistan,' op. cit.

69 *Jang* (an Urdu daily, Rawalpindi), 8 August 1989, reported that twenty-five Soviet transport planes loaded with military supplies were landing every day at Kabul and Mazar-i-Sharif.

70 Author's interview in Peshawar on 2 January 1990 with Pacha Gul Wafadar, former minister of the Kabul regime who defected to the *Hizb-i-Islami* of Hikmatyar.

71 *New York Times,* 16 April 1992.

Chapter 5

The Geneva Accords

PROXIMITY TALKS

The negotiating process that finally resulted in the signing of the Geneva Accords in April 1988 was innovative as well as complex, given the number of actors and the diversity of interests involved in the conflict.[1] A mode of proximity talks was engineered because of the refusal of Pakistan to conduct face to face negotiations with the Marxist regime, which it refused to recognize. Much before the formal negotiations began in June 1982, both Afghanistan and Pakistan had expressed interest in a political settlement of the crisis, though for different ends. While Afghanistan's motive was to seek an end to what it called 'intervention' from Pakistan, the latter for geopolitical reasons and security considerations wanted the withdrawal of the Soviet forces and the establishment of a broad-based government that would enable the millions of refugees on its land to return to their country. The basic elements of such a settlement existed in the United Nations General Assembly resolution of 20 November 1980, which called for: (a) the preservation of the sovereignty, territorial integrity, political independence and non-aligned character of Afghanistan; (b) the right of the Afghan people to determine their own form of government and to choose their economic, political and social system free from outside intervention, subversion, coercion or constraint of any kind whatsoever; (c) the immediate withdrawal of the foreign troops from Afghanistan; (d) the creation of the necessary conditions which would enable the Afghan refugees to return voluntarily to their homes in safety and honour.[2]

On the basis of these essentials and the authorization from the General Assembly to negotiate a political solution of the crisis, the

Secretary General appointed Diego Cordovez to act as an intermediary between Islamabad and Kabul. Not only did the fundamental interests of Afghanistan and Pakistan diverge on the issues of foreign troops and intervention, but also a climate of cold war had set in between the Soviet Union and the United States, which made the diplomatic task of Mr Cordovez one of the most delicate and certainly difficult. On account of his diplomatic genius, generous skills, perseverance and, more importantly, unfailing optimism, the negotiations progressed steadily. Two sets of factors constantly interfered with the United Nations' diplomacy. First, the involvement of the superpowers in the Afghan conflict. This was complicated by the military presence of the Soviet Union in Afghanistan and its engagement in counter-insurgency operations. The geopolitical characterization of the Soviet intervention by the United States was fundamental to the shaping of its response in strengthening the Afghan resistance. Second, partly influenced by the Afghan crisis, the rivalry and mistrust between the US and the Soviet Union erected strong barriers in the way of a political settlement. Largely due to this factor, the UN mediator found it extremely difficult to break the diplomatic impasse at the critical stages of the Geneva talks. Many of the constraints lay in the fact that neither Kabul nor Islamabad acted independently of the interests and preferences of their superpower supporters.

It was only after shuttling from capital to capital for several months that Cordovez succeeded in bringing the representatives of Kabul and Islamabad to Geneva in June 1982. Iran, another regional state affected by the Afghan crisis, refused to participate in the negotiations citing exclusion of the Afghan resistance, which it thought was a legitimate party to any solution. At this point the superpowers and their regional allies were cautious and sceptical about the efficacy of the UN diplomacy. Washington believed that the Soviet Union wanted to use the Geneva talks only to gain legitimacy for the Marxist regime and to head off criticism of its policies at the meetings of the UN General Assembly and the Organization of the Islamic Conference.[3] The Soviet conditions for withdrawing from Afghanistan, such as international recognition of the regime it had installed in Kabul and an end to the arming of its opposition, strengthened these fears. Since, however, no one wanted to appear to be blocking the political settlement, the UN's attempts to inaugurate the talks finally met with success. Engaging the parties in a diplomatic process and sustaining interest in the face of deep differences was itself a remarkable achievement.

The first and foremost task before Cordovez was to obtain agreement on format and agenda. Before convening the meeting he secured the understanding of Islamabad and Kabul on these issues. Pakistan's insistence on accepting the right of the Afghans to elect their government as an agenda item had stalemated UN diplomacy. Cordovez was able to convince the Pakistanis that under no circumstances would the Kabul-Soviet side agree to it. The question he often raised with Pakistani representatives was how to effectuate an independent process of self-determination against the will of a sitting member of the United Nations. Pakistan gave up on this point but impressed upon Cordovez the need for a broad-based, representative government in Kabul. This he simply conveyed to the Soviet leadership without obtaining any commitments. Much of the ground was paved before the actual talks started in Geneva. The later meetings were used to formalize the understandings reached separately with the various parties. During the discussions, Cordovez presented a concept of comprehensive political settlement that would address the fundamental concerns of the two parties. He argued that all the elements of the settlement, though negotiated separately and one at a time, would be interdependent. On the face of it, it looked simple, but in essence it was complex, as interconnectedness would depend on the final resolution of all major and minor problems. Separating the less knotty issues from the most stubborn ones while retaining them all on the agenda was part of the piecemeal negotiating strategy of Mr Cordovez. Four issues figured prominently on the agenda from the outset, which Pakistan and Afghanistan agreed to discuss. These were: withdrawal of foreign forces, mutual non-interference, international guarantees and the return of the refugees.[4]

Substantial progress was achieved during the second round of the Geneva talks held between 11 and 22 April, 1983. Agreement was reached in principle on the withdrawal of Soviet forces, non-interference and the return of refugees. A number of factors contributed to this breakthrough. First and most important was the leadership change in the Soviet Union. In October 1982, six months before the second round, Yuri Andropov, who as the KGB chief had reportedly opposed military intervention in Afghanistan, took over power. During Brezhnev's funeral in November 1982 he told General Ziaul Haq in unequivocal terms that Moscow was serious in withdrawing from Afghanistan. Second, Pakistan appeared to be accommodating as it believed that the Soviet Union could not be forced out merely through military means. After obtaining direct and clear affirmations from the Soviet leadership, Pakistan's interest in the Geneva talks strengthened. Equally strong signals of withdrawal were conveyed by Andropov to Perez de Cuellar and Cordovez when they met him in March 1983. Inspired by these hopeful signs Cordovez indicated after the second round that ninety-five per cent of the draft agreement had already been drawn up.[5] The remaining five per cent was related to the guarantees of non-intervention that were to be furnished by the US and the Soviet Union. Moscow had shown its willingness to sign the accords as a guarantor when in final shape.

The optimism raised by the flurry of activities and positive statements from Moscow and Islamabad evaporated soon after Pakistan's Foreign Minister Sahibzada Yakoob Khan met US Secretary of State George Shultz in Washington on 25 May 1983, about two weeks before the start of the third round of the Geneva peace talks. As it transpired, Washington did not share Pakistan's view of Soviet intentions. Apparently Washington was opposed to any deal on Afghanistan that would not explicitly allow the Afghan resistance to replace the Soviet-backed regime. While Pakistan seemed prepared to push forward the Geneva process, a number of issues remained unresolved, among them a time-frame for the Soviet troop pull-out, the linkage of the withdrawal with the cessation of aid to the resistance, international guarantees and the replacement of the Kabul regime.

The Soviets were unwilling to discuss the total replacement of the Kabul regime. However, as Pakistan had insisted that Babrak Karmal was not acceptable, the Soviets gave indications that he would be replaced if other elements of the accord were agreed upon. A little while before the third round of the Geneva talks, Moscow for the first time publicly declared that it was willing to give a timetable for the total withdrawal of its forces from Afghanistan.[6]

Many observers believe that a political settlement of the Afghan crisis was possible in the middle of 1983.[7] This opportunity was lost because Pakistan bowed to US pressure.[8] The US policy on Afghanistan was primarily influenced at that time by the hard-liners who suspected that the Soviet Union was not sincere in abandoning Afghanistan. This view suggested that Moscow was using the Geneva talks for achieving two purposes: buying time for consolidating its control and deflecting international criticism of its war against the Afghan people. A number of State Department officials who were always ready to abandon the resistance favoured a settlement with the Soviet Union. They found themselves in the minority and totally ineffectual due to the anti-communist passion of the Reagan administration.[9] When Pakistan attempted to nudge Washington into supporting the political settlement it found considerable sympathy in this group, but it was too weak to effect a policy change. Rather, Pakistan was told point blank that Washington would not support what it called a 'flaky' settlement that did not include the participation of the Mujahideen in a future political set up in Kabul.[10] Unfortunately, the domestic political interests of the Zia regime were too tied up with the US economic and military aid package to allow it to act independently on the Afghan issue. Only at the risk of political instability and estrangement from Washington could Pakistan procure a political settlement. Moreover, a complete policy consensus never existed in the dominant elite structure of Pakistan. The point is that even if such a consensus had been available, Pakistan was not willing to sacrifice the strategic advantages of its security links with the United States by striking an independent deal with Moscow.

The political atmosphere of the Geneva talks deteriorated after the June 1983 session. Moscow began to accuse Pakistan of reversing its earlier undertaking by raising afresh the issue of an exclusive Mujahideen government in Kabul as a precondition for signing the accord. On the other hand, Pakistan felt that with the worsening physical condition of Andropov, the conservatives in Moscow were dragging their feet on the issue of the timetable. After his death, the shift in Soviet policy became quite visible. By gaining control of the policy process, the hard-liners side-tracked by Andropov began to insist on direct talks between Kabul and Islamabad to produce a bilateral non-interference treaty. Kabul seemed to be encouraged by the political change in Moscow. Karmal now vehemently rejected a proposal put forward by Cordovez for the Soviet withdrawal to start simultaneously with the aid cut-off to the resistance and before recognition of his regime by Pakistan.[11]

It was only after more than a year that the next round of the Geneva talks could be held in August 1984. Once again Pakistan demonstrated some degree of independence from Washington by agreeing to the timing of aid termination to the Mujahideen. However, the Kabul-Soviet side showed no flexibility in their earlier position on this issue. Meantime, the US Congress began to take a more hawkish attitude towards the war in Afghanistan. Popular support for the Afghan Mujahideen was far greater than for any other anti-communist insurgencies in the Third World; consequently, the Congressional allocation for the Afghan resistance substantially increased. Another factor that derailed the Geneva peace talks was the escalation of the Soviet counterinsurgency campaigns against the Afghan Mujahideen in 1985. The new Soviet leader, Mikhail Gorbachev, was ill-advised by his military establishment that a more aggressive posture against Pakistan, coupled with the intensification of military strikes against the guerrillas, would turn the tide in favour of the Kabul regime. The US and Pakistan, however, responded to Soviet military efforts by upgrading the supply of weapons to the resistance, in terms of both quality and quantity. This spiral of escalation in the Afghan conflict brought back the issue of Soviet motives, dampening hopes for an early solution.

A number of developments later put the Geneva process firmly on the rails again. The first was the resumption of US-Soviet formal talks on Afghanistan in June 1985, which had been suspended since July 1982.[12] A key shift was seen in the White House's abandonment of its policy of 'bleeding' the Soviet Union in Afghanistan, and its agreement to support and take an active role in the UN peace process. The US Secretary of State, George Shultz, showed willingness to accept the installation of a non-aligned government which would pose no threat to the Soviet Union. In return he demanded that the Soviet Union agree to a timetable for the withdrawal of its troops from Afghanistan.[13] The second development was a series of positive indications from the Soviet Union in 1986 that it was serious about ending its military presence in Afghanistan. A decision by the Central Committee of the Soviet Communist Party not to commit more troops in Afghanistan, even with the intensification of the conflict, marked a significant change.[14] It was a bold snub to the Soviet military command that had evidently requested more troops. A more pronounced shift began to occur when Mikhail Gorbachev termed the Afghan conflict a 'bleeding wound' in his speech to the 27th Congress in February 1986. More than a public acknowledgment of the problem the war posed for the Soviet Union, this was intended to weaken the voices in the Soviet establishment which were advocating the exercise of the military option in Afghanistan.

Although Mr Cordovez frequently cited the fact that three documents—on mutual guarantees of non-interference and non-intervention, international guarantees, and the repatriation of refugees from Pakistan—were virtually complete by 1986, some knotty issues remained to be resolved.[15] The timetable for the withdrawal of the Soviet forces and some method of its verification were considered by Pakistan to be too important to be separated from the rest of the agenda. The impasse on the verification issue was to some extent resolved when a tentative understanding on a UN role in monitoring the troop pull-out was reached during the session of February-March 1987.[16] A compromise on the operational interrelationships between the different constituent elements of the accords was also worked out. Three issues: the timetable of the troops' withdrawal, the political arrangements that would replace the Kabul regime, and the linkages between the Soviet withdrawal and the termination of aid to the Afghan guerrillas, remained still unresolved.

When Kabul finally offered a timetable of twenty-two months in March 1987, it was rejected by Pakistan as being far longer than the logistical requirement of pulling out troops. Pakistan suspected that the Soviet Union wanted to build up a puppet regime that would remain in control after the withdrawal of its forces. Another reason why Pakistan demanded a short time-frame was its concern that the Soviets would use the long withdrawal period to crush the Afghan Mujahideen. But after some hard bargaining, the Kabul representatives scaled down the period to eighteen months. Although this was still not acceptable to Pakistan, which insisted on a single-digit time-frame, it showed that a compromise could be reached.

The one-day formula had been discussed for some time to link the beginning of the Soviet withdrawal to a simultaneous cutting-off of aid to the Mujahideen by Pakistan and the United States. Mr Cordovez had obtained Soviet consent on this point. This was partly the reason why the Kabul regime insisted on a longer time-frame, in an effort to gain a decisive military edge over the Mujahideen. Fearing that this move might cause damage to the guerrillas, Pakistan and the United States accepted the idea of the one-day formula in 1987, on condition that the Soviet withdrawal was 'front-loaded', i.e. the bulk of the Soviet forces would be disengaged and sent home in the early phase.

The most intricate problem however related to the future political arrangements in Kabul. Here the differences remained unbridged, and this issue was left out of the Geneva Accords. Moscow called for a government of national reconciliation, meaning a broad-based government comprising the existing structure of the Kabul regime and elements from its opposition. This idea was motivated by the Soviet desire on the one hand to assure the survival of the Kabul regime, and on the other to cultivate some groups from among the opposition which would accept the primacy of its interests in return for concessions such as autonomy or even fundamental change in the ideological character of the regime. Other concerns related to the political vacuum which Moscow feared its pull-out would produce and which could precipitate another round of bloody civil war. Washington and Islamabad did not accept the idea of a coalition government, insisting that the withdrawal should not await these arrangements. First they feared that, as the formation of a coalition government would be a complex, if not impossible, task, the Soviets would use it as a pretext for delaying their troops' pull-out. Also, their clients, the Mujahideen, appeared utterly opposed to any principle of power sharing with the Kabul regime. Second was a gross miscalculation on the part of the United States and Pakistan, that with the withdrawal of the Soviet forces the Kabul regime would collapse. In hindsight, it appears that the offer of a coalition government merited serious consideration when Moscow was quite willing in 1987 to accept a non-communist to head such a government. It was after persistent refusal by Islamabad and Washington to negotiate on this rather crucial aspect of the peace process that the Soviet Union finally abandoned this scheme. But they firmly told the new Afghan ruler, Dr Najibullah, to vigorously pursue national reconciliation.[17] This programme focused on entering into agreements with influential tribes, religious leaders, and internal commanders. The Marxist regime offered autonomy and power-sharing along with the usual incentives of patronage to gain some credibility.

The momentum towards a political agreement started rolling faster in 1987. The Soviets seemed willing to make major compromises to seek an honourable way out of Afghanistan.[18] In May 1986 they had removed Babrak Karmal, who was seen by Pakistan as one of the obstacles in the way of an agreement. Moscow replaced him with a more pragmatic leader, Najibullah, who showed greater support for the Soviet withdrawal and appeared sincere in pursuing national reconciliation. In September 1987 the Soviets met the most important Pakistan condition of a shorter time-frame. Much of the ground work was paved before the convening of the final round in April 1988. In order to push Kabul further towards accepting a shorter time-frame, and also to make the deal more acceptable to Pakistan and the United States, Gorbachev categorically stated on 8 February 1988, that the Soviet Union would start withdrawing its forces from the middle of May.[19] He even hinted at a unilateral withdrawal if Afghanistan and Pakistan failed to reach an agreement in Geneva.

With clear signals emerging from the Soviet Union with regard to its withdrawal from Afghanistan, two issues surfaced once again. The first was the demand by the US Secretary of State that an agreement on Afghanistan should include not only a withdrawal of Soviet forces and a halt in American aid to the Mujahideen, but also an end to Soviet military aid to the Kabul regime.[20] The second issue concerned the formation of an interim government to replace the

Kabul regime and had been raised by General Zia of Pakistan.[21] The Pakistani negotiators however, now under the command of Prime Minister Mohammad Khan Junejo, who appeared more eager to resolve the Afghan problem, dropped this demand.[22] Compromising the position on the formation of the interim government produced serious consequences, as we shall examine in the section on implications of the Accords. But the most negative aspect of the Accords was the understanding between the US and the Soviet Union to continue aiding their respective clients in the war. The Soviets cited their treaty obligations and, of course, a political interest in the survival of the Kabul regime, while the United States termed any cutting-off of support for the resistance unfair, given that Moscow had left behind a tremendous amount of lethal supplies.[23] The US administration perhaps wavered on the continuation of military assistance to the Mujahideen, but Congress was adamant on linking the termination of aid to the guerrillas to a similar undertaking on the part of the Soviet Union.[24] This was precisely the position the US Secretary of State took in his correspondence with the Soviet leaders. The United States told the Soviet Union in unambiguous terms that it retained 'the right, consistent with its obligations as guarantor, to provide military assistance to parties in Afghanistan. Should the Soviet Union exercise restraint in providing military assistance to parties in Afghanistan, the US similarly will exercise restraint.'[25] The Soviet response to that letter, though not revealed, is widely believed to have neither explicitly rejected the US position nor accepted it, but the Americans and Pakistanis have interpreted it as an understanding for a positive symmetry, arguing that US military supplies to the Afghan resistance would not constitute a violation of the Geneva Accords. Undoubtedly, no matter what the tacit understanding between Washington and Moscow, legally the so-called symmetry could not hold without violating the clause on non-intervention. The continuation of the conflict points to the fact that by leaving out the critical issues of the interim government, a cease-fire, and the requirement that the superpower 'guarantors' halt the supply of lethal weapons to the warring parties, the Geneva Accords offered very little in terms of resolving the Afghan problem. Given the convoluted logic of *realpolitik,* the Accords were legally rigged so much that, by the time the ink had dried, they had lost relevance to the situation inside and around Afghanistan.

THE CONTENTS

The Geneva Accords comprise four Instruments and a Memorandum of Understanding appended to the package. Two are bilateral agreements between Afghanistan and Pakistan, the third is the US-Soviet Declaration on International Guarantees, and the fourth is an agreement on the interrelationships for the settlement of the situation relating to Afghanistan signed jointly by the parties, with the guarantors as witnesses.

The first of the bilateral agreements between Afghanistan and Pakistan, entitled 'Principles of Mutual Relations, in particular on Non-Interference and Non-Intervention,' sets out thirteen specific obligations. Some of the activities of the contracting parties constituting interference or intervention in the language of this treaty are noteworthy. Article II (4) binds both Afghanistan and Pakistan 'to ensure that its territory is not used in any manner which would violate the sovereignty, political independence, territorial integrity and national unity or disrupt the political, economic and social stability of the other High Contracting Party.' Article II (7) obliges each party 'to refrain from the promotion, encouragement or support, direct or indirect, of rebellious or secessionist activities against the other High Contracting Party, under any pretext whatsoever, or from any other action which seeks to disrupt the unity or to undermine or subvert the political order of the other High Contracting Party.' Article II (8) more clearly commits each signatory 'to prevent within its territory the training, equipping, financing and recruitment of mercenaries from whatever origin for the purpose of hostile activities against the other High Contracting Party, or the sending of such mercenaries into the territory of the other High Contracting Party and accordingly to deny facilities, including financing for the training, equipping and transit of such mercenaries.' Article II (11) requires each party 'to prevent any assistance to or use of or tolerance of terrorist groups, saboteurs or subversive agents against the other

High Contracting Party.' Article II (12) puts each party under obligation 'to prevent within its territory the presence, harbouring in camps and bases or otherwise, organizing, training, financing, equipping and arming of individuals and political, ethnic and any other groups for the purpose of creating subversion, disorder or unrest in the territory of the High Contracting Party and accordingly also to prevent the use of mass media and the transportation of arms, ammunition and equipment by such individuals and groups.'

The reason these clauses have been cited at such length is to show how effectively the Accords shut out every possible means by which Pakistan could continue to provide sanctuary to the Afghan resistance or serve as a conduit for arms supplies. While the treaty set comprehensive limitations on Pakistan not to act in support of the resistance, it categorically excluded Soviet military assistance to the Marxist regime or the presence of Soviet military advisers in Afghanistan. No matter how stringent the legal obligations outlined in the treaty on non-intervention, both Kabul and Islamabad never altered their policies for a single day. The Marxist regime did not stop subversion and terrorist activities such as bomb explosions in urban areas of Pakistan. Nor did Pakistan discontinue its support to the Afghan guerrillas, including the supply of weapons from the United States. Firstly, Pakistan never officially acknowledged any activity that might be defined as interference or intervention under the Geneva Accords, although it was an open secret that most of the supplies to the Afghan guerrillas originated from or were transmitted through Pakistani territory. Secondly, the understanding on 'positive symmetry' between the superpowers, which Washington could not operationalize without using transit points in Pakistan, confounds the legal and political character of the bilateral accord on non-intervention between Kabul and Islamabad.

The second bilateral agreement between Afghanistan and Pakistan related to the voluntary return of refugees. It guaranteed the returning refugees fundamental civil rights and obliged the Kabul regime to take appropriate measures for their implementation. There was a provision for the setting up of mixed commissions for the purpose of orderly and peaceful repatriation of refugees, which, commencing on 15 May 1988, would last for eighteen months. It was envisaged that the process would be completed during that period, but if necessary, further arrangements could be worked out. In fact, not a single Joint Commission was set up to facilitate the return of the refugees. Nor were the conditions inside Afghanistan peaceful enough to attract the refugees back home. Non-recognition of the Marxist regime was another factor which prevented normal interaction between the two neighbours. Since a political settlement was never achieved between the Marxist regime and the Afghan resistance, and sufficient infrastructural support was not available to the returnees, the refugees continued to stay in Iran and Pakistan.

The 'Declaration on International Guarantees' signed by the United States and the Soviet Union committed them 'to invariably refrain from any form of interference and intervention in the internal affairs of the Republic of Afghanistan and the Islamic Republic of Pakistan and to respect the commitments contained in the bilateral Agreement between the Republic of Afghanistan and the Islamic Republic of Pakistan on the Principles of Mutual Relations, in particular on Non-Interference and Non-Intervention.' This declaration, not requiring ratification by either the US Senate or the Presidium of the Supreme Soviet, fell short of a fully binding international treaty. To further weaken the commitments which could be invoked under the traditional notion of some legality attaching to such declarations, the United States submitted a qualifying statement praying to read the declaration in the context of its position on what was then known as 'positive symmetry.' This was devised to create the legal loopholes for the continuation of US military assistance to the Afghan resistance.

The fourth instrument, entitled 'Agreement on the Interrelationships for the Settlement of the Situation relating to Afghanistan', was signed by Afghanistan and Pakistan and, as witnesses, by the United States and the Soviet Union. The purpose of interrelationship of all the Accords to achieve a political settlement was stated somewhat optimistically: 'from 15 May 1988, there would be no interference and intervention in any form in the affairs of the parties; the international guarantees would be in operation; the voluntary return of the refugees to their homeland would start and be completed within the time-frame specified.' The continuation of conflict in

Afghanistan and lack of accommodation between Kabul and Islamabad pointed to the contrary. The very purpose of promoting a political settlement but without involving the real parties to the conflict, namely the Mujahideen and internal commanders, was self-defeating.

The annexed 'Memorandum of Understanding' outlined modalities and logistical arrangements to facilitate the work of a representative of the UN Secretary General, and personnel under his authority, to investigate any alleged violations of any provisions of the Accords.

POLITICAL AND SECURITY IMPLICATIONS

The end of the Soviet interventionary presence in Afghanistan, though a major contribution of the UN-sponsored peace process, did not significantly diminish the severity of internal conflict. The reason was that the Geneva Accords which facilitated the Soviet Union's way out of Afghanistan left the vital issues of conflict largely unresolved. The arrangements under these Accords, essentially between Afghanistan and Pakistan, in no sense constituted a peace settlement as they failed to address the question of self-determination, which was believed to be fundamental to any endeavour to restore peace. Nor did Soviet interest in the stability and survival of the Kabul regime loosen. Moscow left a large arsenal in Afghanistan and continued to supply more weapons, enabling Kabul to use indiscriminate violence against its opposition. Even if no longer motivated by ideological concerns, there were purely pragmatic reasons for Moscow to stabilize the regime headed by Najibullah. The changing political scene in the Central Asian Republics added more than ever before to the Soviet imperative to support a friendly regime in Kabul that would be sensitive to its interests in the adjoining areas.

Compensating for its military presence with dedicated material support to its allies in Afghanistan was a clever move by the Soviet Union, which was not expected to arouse any significant degree of international opposition. Equally important was the consideration that the Soviet Union under Mikhail Gorbachev, concerned with internal restructuring and desiring integration with the Western community, no longer appeared to pose a geopolitical threat to the region. But this did not mean that the Soviet Union was prepared to accept a lesser world role or abandon its Great Power interests by seeking reconciliation with the Western powers. Rather it wanted to capitalize on its new image and promote more those interests which would not be seen as incompatible with the Western vision of the world order, and not seek purely power-related goals. Therefore, diplomatic overtures towards the West did not affect Moscow's support to the Marxist regime in Afghanistan.

The political and security environment of the region around Afghanistan did not improve qualitatively with the signing of the Geneva Accords. None of the parties in the Afghan conflict abandoned hopes of dealing a knockout blow to the others, nor did Moscow and Washington bridge their differences on a political settlement that would be acceptable to their respective clients. Even the disengagement of the superpowers did not immediately yield a solution because the Afghan conflict, though influenced by external powers, had its own dynamics. Our basic argument is that unless national and regional reconciliation was achieved through political means, change in the interrelationship between the superpowers or the legal framework of the Accords could not by itself be a sufficient factor to achieve peace in Afghanistan. With this caveat, an attempt is made to examine the political and security implications of the Accords and of the Soviet withdrawal for the Kabul regime, the resistance and the neighbouring states.

The Soviet withdrawal under the umbrella of the Geneva Accords did not much alter the balance of forces inside Afghanistan. Before pulling out its troops, Moscow made sure that the Kabul forces were well supplied with the military equipment needed to defend themselves effectively. An optimal level of military build-up, and the commitment to continually refurbish Kabul's losses, proved to be adequate compensation for the disengagement of Moscow's troops.[26] In the significant areas of advice, intelligence support and military-supply relations, the Soviet involvement in Afghanistan did not end. Such an involvement was neither a violation of the Geneva Accords nor was it costly in political and diplomatic terms. By ending their military presence yet retaining a firm interest in the defence of a puppet regime, the Soviets were able to kill two birds with one stone. First, Moscow relieved domestic pressures and widespread resentment over the loss of Soviet lives in a foreign war whose objectives were seen at the popular level as ambiguous and the protracted conflict as costly.[27] Gorbachev, after realizing that the war against the Afghan guerrillas was not winnable, decided to extricate Moscow in an honourable way which would cloak military weakness and show that disengagement was the result of a negotiated settlement. The Geneva Accords provided the excuse for the Soviet Union to make a virtue out of necessity. Second, the withdrawal was projected to rehabilitate the Soviet image and to tell the world that it was serious about cleaning up the mess left by Brezhnev and his comrades. Moscow attempted to benefit from this theme by confessing that the invasion of Afghanistan was a violation of Soviet law and constituted a crime.[28] For the world at large, the end of the Soviet occupation generated an erroneous impression that Moscow was out of the war business in Afghanistan.

A close scrutiny of the Soviet strategic interest in Afghanistan would reveal that Moscow was not willing to abandon its tragedy-stricken neighbour even after negotiating its withdrawal. With rumblings of nationalist movements in the former Central Asian republics, and some reports of covert Mujahideen support to the anti-Soviet underground organizations there,[29] Moscow's stake in the stability of a docile client regime in Kabul remained fairly strong. In the past, Central Asian Mujahideen had used sanctuaries in northern Afghanistan to organize attacks against the Soviet occupation forces.[30] Mindful of that, the Soviets wanted to use their influence over Kabul to prevent the infiltration of hostile elements.

Even after the Geneva Accords, the Soviet political presence in Afghanistan did not lessen. There was nothing in the Geneva Accords to curtail the Soviet relationship of influence with the Marxist regime. A large though unspecified number of advisers, in the guise of technical experts engaged in the execution of development projects, continued to exercise control over the various departments and institutions of the Kabul regime.[31] Intelligence agencies and military establishments, trained and fully equipped by the Soviet Union, retained their Soviet 'experts.' The Soviet involvement in these institutions was directly related to the war effort of the Kabul regime. Nor did the role of the Soviet embassy in the factional politics of the ruling group in Kabul decline. The Soviet ambassador in Kabul acted as nothing less than a viceroy. The Soviet diplomats fully participated in the political drama of Afghanistan, ostensibly promoting unity and helping weed out recalcitrant particularists. In times of crisis, frequently caused by the struggle for power, the Soviet embassy was the ultimate refuge for loyal Kabul-regime functionaries. In the March 1990 *coup* attempt led by the former Defence Minister, General Shahnawaz Tanai, Najibullah sought protection in the Soviet embassy shortly before the bombardment of the presidential palace started.[32] Headquartered in the embassy, he was able to instruct the loyalist forces to strafe the Ministry of Defence, thereby cutting the rebel general's communications with his troops, which enabled Najibullah to survive the coup.[33] Needless to say, Soviet military supplies to restock Kabul's losses began arriving as soon as the forces loyal to Najibullah had regained control of the airports.[34]

Compared with the Soviet resolve to protect its clients in Kabul, the United States was less inclined to maintain the same level of support that it did when the Soviet forces were in Afghanistan. There was a suspicion in Pakistan and among the Afghan resistance groups that, in the absence of the Soviet factor, Washington's interest and involvement in the Afghan conflict might decline further. The United States, however, honoured its previous commitments, both to save credibility and as a moral debt to the Afghan resistance. For a number of reasons, though, its Afghan policy, which was already softening, changed entirely with the break-up of the Soviet Union and the transformation of Russia that followed the failure of the August 1991 coup attempt by the communist hard-liners. First, Washington's strategic interests in Afghanistan, which were never seen as political, were achieved with the Soviet troop withdrawal. Second, in the changing global atmosphere aided by the fall of communism in the former Soviet Union and Eastern Europe, it was not possible for the US administration to achieve consensus on assisting a conflict that seemed to defy political solutions. Finally, the Islamist orientation of some of the resistance groups emerged as more worrisome and was one of the reasons why the US was no longer enthusiastic about a military victory by the Mujahideen.[35] The US feared that such a victory would bring into power doctrinaire Islamists who saw in the US no less an evil than the former Soviet Union. Therefore, from 1991 Washington began supporting the moderate Mujahideen factions more assertively because they appeared more willing to negotiate a peaceful political transition than the hard-liners.

With the changes in the global power relations, one could discern signs of a shift in US policy on two important issues. First, Washington no longer stuck to the condition that Najibullah step down before a transitional government was formed, to facilitate an Afghan dialogue on political reconciliation.[36] Second, there was greater emphasis on a broad-based coalition in which some elements of the Marxist regime could also be inducted. The structuring of such an arrangement as a part of the Geneva talks would have been favourable to the resistance. It might have saved the thousands of Afghan lives which were lost in the ensuing battles to topple the Marxist regime.

In the final stages of the negotiations Pakistan, apparently under US influence,[37] compromised its long-standing position on the formation of an interim government before the departure of the Soviet troops.[38] Without such a concession, what would be the significance of a bilateral treaty between Kabul and Islamabad for the Afghan resistance that was never a party to it, and that expected to gain nothing from the Soviet withdrawal? Many of the resistance leaders argued that the Soviet Union was disengaging itself anyway. Gorbachev's February 1988 statement that, Accord or no Accord, the Soviet Union would start withdrawing its troops on 15 May was a clear indication, much in line with earlier assessments, that the stalemate had crushed the Soviet resolve to continue using its troops. Had self-determination, whatever its mechanism in the objective conditions of troubled Afghanistan, been made a central issue in the Geneva Accords, many channels of national reconciliation might have opened up. Leaving the political settlement out of the deal, and allowing the Soviet Union to withdraw its troops while remaining indirectly involved, kept the military stalemate unchanged.

The Soviet withdrawal undoubtedly had a benign effect in political and psychological terms on the efforts of the Kabul regime to gain domestic acceptance and some international respect. Firstly, the stigma of Soviet involvement, which had for a long time incapacitated the political moves of the Kabul regime, was removed. Secondly, one of the fundamental causes of the national uprising that motivated the Afghan Jihad disappeared with the Soviet troop pull-out. Finally, Najibullah's initiatives to promote national reconciliation began to earn credibility as he demonstrated flexibility and struck local deals with some tribal chiefs and certain internal commanders.[39] Also, he showed some inclination to change the basic character of the regime and his party.[40] This was a strategy that aimed at isolating the Afghan resistance by fostering war-weariness among the population on the one hand, and on the other entering into deals with local influentials which offered autonomy in exchange for cease-fires. Far deadlier, though, was the propaganda that the Mujahideen could not win the war. Whatever partial success was achieved through a mixture of propaganda and inducement was eliminated by the Tanai *coup*. Rifts within the ruling party remained deep and the Afghan armed forces never emerged as a homogeneous and disciplined group. A number of mutinies in the post-Soviet withdrawal period were testimony to this fact.[41] Encouraged by the Tanai *coup*, some of the resistance groups and internal commanders began giving serious thought to striking an alliance with some elements of the armed forces to topple Najibullah.[42]

As the end of the Soviet occupation of Afghanistan began to change international attitude towards the conflict, Pakistan started to reassess its role and options. With dwindling international support, and due to changes in US policy, Pakistan shifted its policy more to the promotion of a political settlement. Islamabad also opened up diplomatic channels with Moscow to bridge the differences on the formation of a broad-based transitional government. It was a pragmatic approach, intended to reduce threats to its security in a fast-changing world.

The lifting of the military occupation of Afghanistan removed a serious obstacle in the way of improving relations between Tehran and Moscow. Coinciding with the completion of the withdrawal from Afghanistan in February 1989, the Soviet Foreign Minister, Edvard Shevardnadze, visited Iran where he signed a protocol to regulate political contacts and to boost cultural and scientific exchanges. Ayatollah Khomeini had his first and last meeting with the foreign minister of any country. He welcomed the Soviet withdrawal from Afghanistan and called for co-operation with Moscow against the West.[43] This is something that also figured in Khomeini's will, but he did not live long enough to see the disintegration of the Soviet Union and the emergence of a new world order vastly favourable to the West.

These changes had a tremendous effect on the political alignments within Afghanistan, resulting in the collapse of the Marxist regime in April 1992. Although a variety of complex factors were at play, the alliance between disaffected military generals and the guerrilla commander, Ahmad Shah Masud, gave it a knock-out blow. This smacked of a *coup,* more or less from within, but with a significant difference: the broader coalition which brought about the change was dominated by non-Pushtun actors. Nonetheless, it proved two points: first, that the Marxist regime could not survive without sufficient economic and military assistance from the former Soviet Union, and second, that the ruling Marxist group was no less divided than the Afghan resistance.

NOTES

1 On the Geneva parleys, *see,* an excellent study by Riaz M Khan, *Untying the Afghan Knot: Negotiating Soviet Withdrawal* (Durham and London: Duke University Press, 1991).

2 Resolutions and Decisions adopted by the General Assembly during its thirty-fifth session, General Assembly Official Records, Thirty-Fifth Session, Supplement No. 48 (A/35/48), (New York: United Nations, 1991), pp. 17-18.

3 *Los Angeles Times,* 25 July 1982.

4 'Progress at Geneva' (editorial), *Dawn,* 25 April 1983.

5 Cordovez's interview in *Dawn,* 22 August 1985. *See also,* Lawrence Lifschultz, 'An Accord in the Offing,' *Far Eastern Economic Review,* 9 June 1983, p. 28.

6 *See,* a statement by Soviet Ambassador to Pakistan Vitaly Smirnov, *Muslim,* 20 May 1983.

7 *See,* for instance, Selig S Harrison, 'A Breakthrough in Afghanistan,' *Foreign Policy,* No. 5, Summer 1983, pp. 3-26.

8 Mushahid Hussain, 'Will America Let Pakistan Settle Afghan Question,' *Muslim,* 10 January 1987.

9 Author's interviews with officials of Pakistan Foreign Office, Islamabad, January 1990.

10 Ted Morello, 'A Hiccup at the Brink,' *Far Eastern Economic Review,* 9 June 1983, p. 30.

11 Selig S Harrison, 'Inside the Afghan Talks,' *Viewpoint* (Lahore), 19 January 1989, pp. 9-10.

12 *San Francisco Chronicle,* 3 June 1985.

13 *Dawn,* 13 August 1985.

14 Maleeha Lodhi, 'Pakistan on the Threshold of a Shift in Afghan Policy,' *Muslim,* 3 April 1986.

15 *Muslim,* 20 April 1986.

16 *Muslim,* 18 March 987.

17 *Dawn,* 2 July 1987.

18 *The Guardian,* 12 January 1988, p. 9.

19 Martin Walker, 'Gorbachev Clears Way for Afghanistan Withdrawal,' *Guardian Weekly,* 14 February 1988, p. 7.

20 *The Frontier Post,* 10 January 1988.

21 *Dawn,* 12 April 1988.

22 Ziaul Haq, the President of Pakistan and strongman of the military establishment, had been opposed to the terms of the Geneva Accords as negotiated by Prime Minister Junejo. According to one of the closest political allies of Ziaul Haq, former ambassador to Afghanistan, Mohammad Aslam Khatak, only in May 1988 did Zia reluctantly and 'with bleeding heart' agree to a political settlement. The signing of the Accords was one of the issues which resulted in the removal of Junejo, which deepened the political crisis in Pakistan. *Muslim,* 17 August 1991.

23 *The Guardian Weekly,* 26 February 1989, p. 19.

24 The Senate unanimously passed a resolution calling for the continuation of US aid of all kinds to the Afghan resistance as long as Soviet aid was going to the Kabul regime. *Congressional Record,* Vol. 134, No. 21, 29 February 1988, p. S1608.

25 *Department of State Bulletin,* Vol. 32, No. 2135, June 1988, p. 55.

26 Rough estimates suggest that Moscow was shipping $300 million worth of weapons to Kabul each month. *Washington Post,* 12 March 1990, p. A10.

27 The Soviet Premier N Ryzhkov put the economic cost of the war at 5 billion roubles (US $758 million a year). *Far Eastern Economic Review,* 13 July 1989, p. 17.

28 *See,* the statement of the President of Council of Soviets and Non-Voting Member of the Politburo. *New York Times,* 28 October 1989, p. 6.

29 *Tass,* reported on Radio Free Europe/Radio Liberty Daily Report No. 48, 8 March 1990, p. 10.

30 Ludwig W Adamec, *Afghanistan's Foreign Affairs to the Mid-Twentieth Century: Relations with the USSR, Germany and Britain* (Tucson, Arizona: The University of Arizona Press, 1974), p. 69.

31 White House statement in *Wall Street Journal,* 11 October 1989, p. A1.

32 Radio Islamabad, 7 March 1990, in FBIS, 7 March 1990, p. 35.

33 Salamat Ali, 'Factional lines blur following the attempted *coup*', *Far Eastern Economic Review,* 22 March 1990, p. 19.

34 BBC report quoted in RFE/RL Daily Report No. 49, 9 March 1990, p. 17.

35 *See,* Representative Lee H Hamilton's comments in 'Negotiate an End to Afghanistan's War,' *Christian Science Monitor,* 14 March 1990, p. 18.

36 *New York Times,* 24 February 1990, p. 8.

37 For this assessment see Rosanne Klass, op. cit., p. 936.

38 Maleeha Lodhi, 'Pakistan's ideas on interim government communicated to Cordovez,' *Muslim,* 27 January 1988.

39 *The Times,* 31 January 1990.

40 After the March *coup* attempt by General Tanai, Najibullah asked his party to end its monopoly of power and to change the constitution and name of the party. *New York Times,* 19 March 1990, p. A8.

41 The Afghan secret police in December 1989 arrested 127 suspects from the armed forces who were involved in a *coup* plot. *Muslim,* 1 January 1990.

42 *New York Times,* 12 March 1990.

43 *Dawn,* 28 February 1989.

Chapter 6

The Marxist Regime: Dependent Survival and Collapse

Building a political and administrative structure, along with making the regime acceptable, was as important in the Soviet strategy for the pacification of Afghanistan as strengthening coercive institutions or attempting to cripple the Mujahideen resistance. The Soviets wished their Afghan comrades to pull their own weight by moderating policies on reforms and religion, and waging most of the war against the resistance. Broadening the support base of the regime was considered by the Soviet advisers as fundamental to effective political control. Power-sharing with prominent political groups outside the Party was devised both to attract domestic support and to isolate the resistance. The Soviets were firm on counselling the PDPA factions to unite and give up infighting—a goal which met with little success. The restructuring of education and the raising of a more disciplined and indoctrinated Party cadre were two areas that were considered vital to the political organization strategy.

In the first place, the Soviets never regarded the war as a separate task from that of overhauling and rebuilding a political order around the Kabul regime. Secondly, they realized, in the face of factionalism, corruption, indiscipline and the narrow support-base of the Marxist regime, that no amount of violence would salvage the situation. They therefore attempted to promote an integrative political strategy combining restructuring, coercion, and co-option. They advised their Afghan clients to reduce their rhetoric on revolutionary changes and to bring the regime closer to popular aspirations by making it efficient, workable and more representative in character.

To appease the inflamed population, the Soviets advised Karmal to backtrack over the issues of reforms and religion. Karmal, after being installed in power, began to project a non-Marxist image by emphasizing that the goal of his Party was not socialism but 'the social development of society.'[1] Equally important was his assertion that he favoured a broad front of all national democratic forces,[2] a stance which aroused hostility towards him within the Khalq faction of the PDPA. He induced non-Party technocrats and influential Afghans to join his regime. To make the shift in the ideological goals of the regime credible, Karmal in 1981 announced changes in the land reforms law exempting the lands owned by *mullahs* (priests), local chiefs and tribal leaders, besides military officers.[3] The Karmal regime carefully avoided making attacks on religion. Attempts were made to get the backing of the *ulema*. A new constitution was promised that would a) respect and adhere to the sacred religion of Islam, b) guarantee private ownership, and c) observe all the political, economic and social rights of the people.[4] To give a humane face to the regime, Karmal also released a large number of political prisoners.[5]

Building a unified Party with a vanguard role in Afghan society was the ambitious goal which the Soviets tried to achieve. Personal rivalries within the PDPA were compounded by social, class and ethnic differences. The installation of Karmal in power with the idea of reintegrating the more numerous Khalq cadre and leadership into the Party was not an intelligent move. The *Khalqis* were as much opposed to the Soviets as to the dominant role of Parcham. Fighting and intrigue against each other continued.[6] It seems that a unified PDPA above ethnic and tribal considerations was an unrealistic expectation. Although Najibullah, as a Pushtun, was better positioned to attract support from Pushtun areas, he was no less opposed to sharing power with the *Khalqis* than Karmal. Khalq's earlier assessment of Najibullah later proved to be correct as distrust, victimization and exclusion of this faction from top government positions continued when he assumed power in 1986. The theme of unity within the Party remained on the formal agenda, but the deep-rooted ethnic and political rivalries prevented a genuine reconciliation between the two wings of the Party.[7]

The Party cadres were neither trained nor disciplined to handle responsibly the complex issue of restructuring a society torn by violence and national uprising. The Party leaders, quite in line with tribal traditions and the spoils system of the Third World, displayed excessive greed in plundering state resources. This is evident from Karmal's unceasing attacks against the 'illegitimate use of state positions' for acquiring state property and taking bribes.[8] Embezzlement, wasteful expenditure, misuse of funds, ambition for office and oppression figured constantly in his speeches on the need to reform the Party,[9] but nothing changed on the ground despite insistent Soviet prodding and the launching of a programme of training Party workers. Privileges and benefits attracted more attention than either serious Party work or addressing the needs of the people. The behaviour of the Party leaders and workers deeply disappointed the followers, and did a lot of harm to the efforts aimed at increasing popular support. Even though the top leaders grew more cognizant of the corrupt practices within the Party, they demonstrated little skill or political capacity to tackle them efficiently. In the Third World environment, corruption of politicians and bureaucrats attracts a lot of attention as a public issue but persists due to lack of accountability, and it may never go away, no matter what the amount of rhetoric, because the institutional bases are weak and civil society has yet to assert its supremacy. In Afghanistan, as elsewhere, the Party's monopoly over state power and a coercive totalitarian control strengthened the arbitrariness of its functionaries, thus breeding more corruption.

Expanding and strengthening the PDPA acquired greater significance under the leadership of Karmal than it did under Amin. In the new thinking, the Party was conceived of not only as an instrument for countering the rising power of the Mujahideen groups but also as a means to generate support for the regime and serve as a link with society. In 1985, the Party claimed that it had more than 140,000 members.[10] At the time of the revolution the membership of the Party was no more than a few thousand.[11] Although the official claims of Party membership were somewhat exaggerated to show the success of the political programme and to win propaganda points, the Party expanded considerably using official resources of patronage. Military personnel, government employees, students and even ordinary people were induced to join the Party by offers of job promotions and privileges. Soviet style mass organizations such as the Democratic Youth Organization and the Democratic Women's Organization were created to broaden the base of support for the regime. However, the majority of Afghans who joined the Party or its affiliated organizations did so more due to expediency than from any ideological conviction. Some resistance groups, though, saw an opportunity to infiltrate the expanding, officially-patronized organizations.[12] By keeping a sizable portion of the urban work force on its pay-roll, the Marxist regime generated a natural interest in its survival.[13] Those employed in the public sector and Party-controlled institutions feared rough justice from the Mujahideen for having been affiliated with the PDPA regime.

The regime attempted to create a number of auxiliary front organizations to extend its reach to the population. One of these was the National Fatherland Front (NFF), which was founded at the end of the PDPA Congress in June 1981.[14] The basic idea was to attract the masses to get organized and participate in pro-government activities. Important groups who for ethnic, religious, or social reasons preferred to stay away from the regime were special targets of the NFF. In 1986 the Party claimed a membership of more than 760,000 in various government-sponsored unions, co-operatives and interest groups.[15] No less exaggerated was the official report which announced the creation of 9,500 NFF councils at local and village level.[16] This was neither likely nor practical given rural migration and a countryside dominated by Mujahideen activities. The operation and effectiveness of the NFF was limited to only a few government–controlled areas.

The creation and expansion of para-military forces under the Ministry of the Interior gained significance in Kabul's strategy to deny the tribal border areas to the Mujahideen. The formation of the militia and the Border Force was also meant to lessen the burden on the Kabul-Soviet forces. It made good political sense to buy the loyalties of tribal chiefs and *maliks* to extend the government's influence, if not control. Money, weapons and official recognition of the semi-autonomous status of the chiefs were promised, in return for their commitment not to allow the Mujahideen the use of their areas. This policy had greater success in the northern areas, where a number of ethnic groups were enticed to form militias. Ironically, however, the Uzbek militia which was raised to fight against the Mujahideen ultimately formed an alliance with them and, in March 1992, turned its guns against Najibullah, thus precipitating the collapse of the Marxist regime. In the Pushtun-dominated areas, many of the militia units could survive only by covertly co-operating with the resistance groups. Those who staunchly supported the government were targeted by the Mujahideen. On balance, the tribal chiefs have been loyal to their local interests and have made a fortune by playing one side against the other.

The Marxist regime's quest for legitimacy focused much on convening *loya jirgas* and seeking support for and endorsement of government policies. This traditional grand assembly of tribes and respected Afghan individuals has historically been convoked to seek a vote of confidence in new rulers. In 1985, Karmal convened the *loya jirga* in which 1,796 delegates participated.[17] Najibullah, who had tribal links in the Pushtun areas, was assigned the task of winning over the chieftains through bribes and offers of government patronage. In order to strengthen the interdiction strategy in the Pakistan border regions, a larger *jirga* of Pushtuns, Baloch and Nuristani tribesmen was held in September 1985. Official sources reported the number of delegates at 3,700.[18] This gathering included quite a large number of tribal leaders from Pakistan and quite a few from Iran.[19] The objectives were a) to enlist the support of local influentials to prevent Mujahideen infiltration, and b) to create a facade of wider acceptance by the tribes.[20] Encouraged by previous successes, the Marxist government in the spring of 1986 invited 1,540 representatives to a *jirga* of what it called 'free Pushtun tribes' from the Pakistani side of the border.[21] Money and weapons were distributed among them to stir up trouble in Pakistan. The delegates were mostly chieftains whose heroin-processing laboratories and smuggling businesses had been destroyed by the Pakistan Army and workers from leftist groups who had been advocating the cause of the Kabul regime in the frontier regions.

Some tribes occasionally succeeded in complicating the situation on one pretext or another. This was partly due to the activities of the Marxist regime's secret police, KHAD (Khidamate Aetilaati Daulati), which employed a large number of agents in the border areas of

Pakistan and was responsible for a variety of acts aimed at generating chaos.[22] The Marxist regime also encouraged Afridi, Shinwari and other tribesmen from the Kurram Agency, North and South Waziristan and Mohmand Agencies on the Pakistani side of the border to hold *jirgas* to oppose Mujahideen groups. Kabul's propaganda centered on how the war and the Mujahideen presence jeopardized the tribes' 'traditional freedom and vital interests.'[23]

No discussion of the Soviet Union's occupation politics would be complete without considering its role in organizing and training the KHAD, which emerged as one of the most important tools of the Marxist regime's internal control. The agency was considerably expanded, upgraded to cabinet level, and renamed the Ministry of State Security a little while before Karmal was removed.[24] The Soviet KGB, which was put in charge of developing the KHAD, practically controlled it. The KHAD was estimated to employ over 30,000 career intelligence staff and regular collaborators.[25] No department of the Marxist regime did better in damaging the resistance than the KHAD. The success of the agency lay in its KGB-type activities focusing on eliminating political dissidents. Inside Afghanistan, the KHAD implanted its agents in educational institutions, trade unions, military units and various ministries. Its employees also operated among ethnic communities and in various localities. A close watch over the population was necessary for ensuring the security of the main cities where the regime had established a relatively stable control.

Besides strengthening the control of the regime, the agency was assigned a variety of tasks which were primarily aimed at sabotaging the resistance. The KHAD concentrated on the following activities: a) infiltrating the Mujahideen organizations and refugee camps, b) conducting sabotage in Pakistan, c) cultivating opposition to the Mujahideen and hatred for the Afghan refugees, d) promoting tribal conflicts and rivalries, e) arresting and interrogating opponents of the regime and f) weeding out dissidents from the army. The KHAD succeeded reasonably well in maintaining discipline in the army and was credited with unearthing a planned *coup* by a large number of military officers in December 1989,[26] and another in March 1990.

KHAD agents did penetrate some of the Mujahideen organizations and not only reported on guerrilla activities but also set one guerrilla group against the other. Not all Mujahideen conflicts can be attributed entirely to the work of the KHAD, but its presence among the Mujahideen certainly exacerbated their rivalries. The Kabul agents were also responsible for the assassination of both internal commanders and dissidents abroad.[27] Such acts served as a double-edged sword. Not only was an opponent eliminated, but the blame was put squarely on the rival Mujahideen organization, thus generating an impression of infighting within the resistance and preventing co-ordination of guerrilla strategies.[28]

To bring about a structural change in the traditional Afghan society, the Soviets attempted to indoctrinate and socialize a new generation through education. A two-pronged educational strategy was adopted. In order to provide disciplined, reliable and efficient future administrators, thousands of young Afghans were sent to the Soviet Union. The targets were young children, often orphans of impressionable age. Estimates of the number of young Afghans sent to the Soviet Union for education vary[29] but reliable sources claim that over 4,000 were sent between 1980 and 1989.[30] This programme was envisaged to educate the Afghans in a different social environment, thus transforming their world view at an early age. As an incentive to go north, the Kabul regime attached prestige to education from the Soviet Union, and the returnees were taken into better governmental positions. There is no doubt that poor and orphan youth educated in the Soviet Union enjoyed better social mobility and perhaps supported the shaping of a new Afghanistan, but the early optimism that Soviet-trained Afghans would become faithful adherents of socialism soon gave way to realism. Afghan youth suffered alienation and discrimination in Soviet society. At a time of new thinking, restructuring and general economic chaos in the former Soviet Union, many came back doubting the efficacy of the socialist path to development in Afghanistan.[31] Even if it could be assumed that a large number of the young people were sufficiently indoctrinated, it was not going to be an easy task to transform the traditional patterns of Afghan society, which was undergoing counter-indoctrination via the liberation mythology of the Mujahideen.

In Afghanistan itself the educational system was reorganized with the objectives of maintaining control over the institutions and orienting the youth towards an uncritical acceptance of the official ideological line. Political control over educational institutions was necessary as they had not only bred hostile ideologies but had also been hotbeds of agitational politics in the past. Half of the Kabul University faculty, whose members were known to hold views opposed to the PDPA, were imprisoned, killed or driven into exile.[32] A large number of Soviet teachers, especially from the Central Asian Republics, were hired to take up positions at the universities and professional schools. Party members were appointed or promoted to higher positions without meeting any eligibility criteria; many of them spied on their colleagues and students.

Another measure intended to transform the educational system was aimed at entirely changing the curriculum. At Kabul University the Faculty of Social Sciences was renamed the Faculty of Letters on the Soviet pattern. Islamic subjects, such as the history of Islamic civilization and Islamic art, were downgraded. A new Russian-style curriculum was introduced: historical materialism, the revolutionary history of workers, history of Russia, new history of Afghanistan, scientific sociology and history of world literature, in which the works of Marx, Lenin and Castro were taught. The new history of Afghanistan was written in the image of the Party by Ghulam Dastagir Panjsheri, then a Politburo member. Most of these subjects were compulsory and were taught by Party activists.[33] Similarly, in all professional institutes and schools, down to primary level, teaching courses were redesigned to emphasize ideology and the Soviet world view. Russian replaced almost all other foreign languages.[34]

The new educational system was also devised to bring about attitudinal changes in the new generation. Elsewhere, indoctrination under controlled conditions did not succeed, as was proved by the revolt of the Chinese students who in 1989 launched a nationwide movement for democracy. They found it impossible to remain loyal to the 'truth' taught to them by the Communist Party. Modern information systems have acquired a global reach, and it would be hard for any type of regime to convert the new generations into loyal adherents of its thoughts and policies. It proved to be more difficult for the Marxist regime to alter the traditional Islamic culture of Afghanistan at a time when democratic revolutions were shaking the foundations of Communist regimes in Eastern Europe and the Soviet population was restive and totally disenchanted with the Communist economy and political rule.

Soviet economic policies in Afghanistan were directed towards two objectives. The first was to recoup the cost of the military equipment sold to the Kabul regime. The only viable source of funds Afghanistan had was its north-eastern gas fields, which had been pumping out natural gas for the Soviet Union long before the invasion. After the intervention the Soviets increased the volume until it accounted for US $300 million-worth of energy supplied to Soviet Central Asia.[35] The Soviets invested heavily in the extraction of mineral resources, mainly to feed their own industries.[36] Defecting Afghan officials constantly accused the former Soviet Union of paying low compensation rates for the gas and other minerals.[35] Worse, 95 per cent of the country's natural gas was supplied to the Soviet industrial centres of Kirghizia and Kazakhstan, leaving the producing country with nothing with which to develop its own industry.

The second objective of Soviet economic planning for Afghanistan was to turn the country into a dependency. Cut off diplomatically by the rest of the world for hosting Moscow's invasion force, Afghanistan became increasingly dependent on its northern neighbour for aid and trade. This was evident from the volume of trade between the two countries. In 1989 Afghan official sources disclosed that trade with the Soviet Union constituted 60 per cent of its total foreign trade, this was expected to increase by 30 per cent by the end of that year.[38]

Aid from the Soviet Union played a vital role in the Afghan economy until the former's own disintegration towards the end of 1991. Since the signing of the treaty of peace and friendship in December 1978, which extended to the year 2000, the Soviet Union had commissioned 170 projects. Another 150 projects were in the process of completion by the end of the 1980s. The importance of these projects can be gauged from the fact that 60 per cent of the total industrial output, 75 per cent of the state sector's industrial production and 60 per cent of the state budget's income were realized from economic enterprises established with Soviet assistance.[39]

FROM KARMAL TO NAJIBULLAH

The Soviets wanted the Marxist regime to increase its political capacity by mending its fences with the people and by changing its rhetoric on Marxist-Leninist ideological themes. The traditional social setting of the country with its Islamic overtones had proved resistant to the Soviet-inspired political change that had fuelled the upheaval. One reason why Moscow had installed Babrak Karmal after the military intervention was that he appeared relatively moderate and pragmatic in dealing with mass unrest. From the outset he was advised by the Soviets to tone down the reforms which he made and to broaden the support-base of the Party and the government through national reconciliation. The objective was to enable the government to gain wider acceptance. Karmal's earlier position at the head of the country's democratic and progressive forces seemed more conciliatory than the sharp ideological dogmatism of Hafizullah Amin.

Despite Karmal's better credentials for normalizing conditions within the country, he was severely handicapped by the circumstances of his assumption of power. From the beginning he suffered from an image problem. He was seen by the Afghans and by the rest of the world as having been brought to power by Soviet tanks. Nothing affronted traditional Afghan honour more than bringing in foreign troops to kill fellow citizens. If the aim was to seek reconciliation with the more numerous and heavily-armed Pushtuns, then Karmal, with roots only in Kabul, was not the right choice. Soviet advisers realized this dilemma when they found tremendous resistance to Karmal from influential Pushtuns within the Party and the armed forces. More importantly, Karmal's successes either in pacifying the country or in rebuilding a unified Party were limited; such tasks required more than his leadership could offer.

In 1985, the Soviets became quite open and frank in portraying the Karmal regime as a failure. The Soviet newspaper *Pravda* on 21 December 1985, for the first time since the invasion, conceded that the Kabul regime had failed to win any significant degree of popular support. The paper admitted that it was a mistake to push through reforms without taking into account social conditions.[38] One could read between the lines that Moscow was running out of patience with the lack of progress towards reconciliation within the country. The *Pravda* comment strengthened speculation that Najibullah was being positioned to replace Karmal. Najibullah's appointment as Party Secretary at the end of the Central Committee Plenum held on 21 November 1985 was a calculated move to help him push his way through the Party apparatus.[41] Through the *Pravda* comment Moscow also conveyed the message that it was serious about negotiating the withdrawal of its troops and promoting a political settlement.[42]

While Najibullah manoeuvred the Politbureau through his control of the KHAD, the Soviets began to distance themselves from Karmal. He was the only head of a ruling communist party to fail to get a private audience with General Secretary Mikhail S Gorbachev during the 27th CPSU Congress of February 1986.[43] In his speech Gorbachev called Afghanistan a 'bloody stump' and indicated his willingness to withdraw troops from that country.[44] This showed his frustration with the slow pace of the return to normalcy in Afghanistan, and was definitely a rebuff to Karmal, who was seen to be advocating the need to engage Soviet forces in the war. Najibullah later claimed that in a meeting in 1985 at which he was present, when Gorbachev proposed pulling out the Soviet forces Karmal said, 'If you go now, next time you will be compelled to send one million soldiers.'[45] Apparently Najibullah did not endorse Karmal's view.

The Soviets gave temporary prominence and recognition to Sultan Ali Keshtmand, the Chairman of the Council of Ministers, by according him an official reception in April 1986. *Izvestia* had earlier published his major article on developments in Afghanistan in which he made a strong argument for a pragmatic approach to resolving the national crisis through political reconciliation.[46] This was much in line with Soviet thinking on Afghanistan at that time, as the heavy reliance on the use of force appeared to be ineffective in gaining control. The Soviets later had second thoughts as Keshtmand belonged to the Shiite Hazara minority, so another leadership change was considered and the Soviets fixed their eyes on Najibullah.

Najibullah, despite his KHAD background, happened to be a better choice. He was a Pushtun, young, energetic and had proved himself an efficient administrator in organizing the KHAD and a capable political tactician in coaxing the tribesmen to support the regime. His political career in Parcham and his Pushtun background[47] were seen as assets in bridging the gap between the Parcham and Khalq factions. On 4 May 1986, the Central Committee of the PDPA elected Najibullah to replace Karmal, his one-time mentor and political guru.[48] The Soviets were careful to manage a peaceful transition. Karmal remained in the Soviet Union for a 'medical check-up' for more than a month, and returned home only two days before the Central Committee decision.[49] The fact that Soviet troops took up positions at key intersections of the capital, while armed Afghan troops guarded all government buildings during the meeting, showed that Karmal was forced to resign.[50] Some *Parchamis* protested Karmal's ouster by instigating student demonstrations in Kabul.[51] Karmal continued in the lesser position of President, hoping to be accepted as an elder statesman, but was phased out when Najibullah was elected President by the *loya jirga* held in Kabul on 29-30 November 1987. Abdur Rahim Hatef, a close associate of Najibullah and president of the National Fatherland Front, listed forty qualifications for a future president, ranging from knowing Dari and Pushtu and being a true follower of Islam to harbouring no self-interest and having decision-making capabilities.[52] The *jirga* had no trouble finding all the required qualities in the person of Najibullah.

STRATEGY FOR SURVIVAL

Najibullah came to power with a mandate to push forward the programme of national reconciliation and to broaden support for the government. As a first step he decided to deal with the image problem his predecessors had suffered from. He played down the old convictions by asserting that he did not want to build a communist society[53]; rather, he stressed the common Islamic heritage of all Afghans and regularly attended Friday congregation prayers in mosques. The religiosity of the PDPA leaders was highlighted in the Press.[54] To win the country's Islamic constituency away from the Mujahideen, Najibullah carefully cultivated links with the *mullahs* by putting large numbers of them on the government payroll.[55] One should, however, be aware that in Islamic societies 'official' *mullahs* have little credibility.

More important were Najibullah's attempts to broaden the support base of his regime. He tried a combination of political moves that appeared to produce some results. He was able to co-opt Dr Hasan Sharq, a non-Party political figure, as the Prime Minister in 1988. The Sharq Cabinet included other non-Party ministers such as Pacha Gul Wafadar, who in December 1989 defected to the Hizb-i-Islami of Hikmatyar.[56] Although Najibullah's partnership with Sharq did not last very long, it showed flexibility on his part.[57] From 1989 onwards he routinely called for the formation of a coalition government of several parties, the drawing up of a new constitution and the holding of free elections under the supervision of an international commission[58] although he continued rejecting the imposition of any preconditions, such as his stepping down before the elections were held. His openess to intra-Afghan dialogue, acceptance of UN mediation[59] and, above all, a secular image made him appear more acceptable to the West than some of the radical Islamic groups who had been the cutting edge of the resistance against the Soviet interventionary forces, and, supported by the US, the linchpin of the Western powers.

Much shaken by the March 1990 *coup* attempt, Najibullah increasingly advocated ending his party's monopoly of power.[60] Also, partly influenced by the restructuring in the Soviet Union, the Afghan leader seemed willing to allow other parties to function, and to co-opt non-Party members into his government. To make this policy credible, out of a total of thirty-six ministers, twenty-three were non-Party members..[61] While this claim might be disputed, there is no doubt that the regime was more flexible in offering government positions to influential individuals. To make his intentions more credible, he purged some Marxist hard-liners from the Politburo and Central Committee.[62]

Najibullah's conciliatory peace overtures towards the opposition in the last years of his rule focused on a ceasefire and negotiations to end the civil war. He approached Zahir Shah and some of the resistance leaders to hold direct talks with his government. His efforts in this respect succeeded when the representatives of Zahir Shah and some moderate parties met him in Geneva in November 1990.[63] These were some of the groups which the Kabul authorities hoped to wean away from the radical Islamic parties to build a new consensus. Shiite parties based in Tehran were another target. Since the Soviet withdrawal, Iranian leaders had shown no enthusiasm for a military solution, which in their estimation would, if sucessful, have brought into power the Sunni variety of Islamic radicals. Thus Iran was inclined to accept Najibullah in the transitional period while bargaining for greater autonomy for the Shiite-dominated areas. Tehran reportedly brought increasing pressure to bear on these parties to accept Najibullah as coalition partner in 1990.[64] It was certainly with the blessing of Tehran that the Afghan President, on his way back from Geneva in November 1990, dashed to Meshhad to confer with some of the Mujahideen leaders.

When the Soviet forces completed their withdrawal from Afghanistan in February 1989, it was widely speculated that the government of Najibullah would either disintegrate or be quickly forced out of power by the Mujahideen. This did not occur. Instead, the departure of Soviet troops allowed Kabul's rulers to present themselves as patriots defending Afghanistan's integrity.

Although the Mujahideen became more aggressive after the Soviet troops left—quickly establishing an interim government and launching a major attack on the strategically important city of Jalalabad—they were unable to make concrete gains. Despite initial set-backs, the regime retained control of Jalalabad, the loss of which would have considerably changed the military balance in the country. Moreover, a series of *coup* attempts, which had been anticipated because of divisions in the armed forces, failed to dislodge Najibullah. This undoubtedly surprised the President's supporters as well as the foreign powers backing the Mujahideen.

How did Najibullah's regime, in place since 1986, manage to survive the pull-out of his patrons? One reason was disunity in the ranks of the Afghan resistance. Najibullah wasted no time in exploiting the resistance movement's lack of cohesion. The government was able to project an image of itself as the only force in the country capable of maintaining order. While this claim might be disputed, the Mujahideen were not able to show to the rest of the world that they were an acceptable alternative. Infighting among the resistance continued to raise the spectre of another round of civil war until an acceptable mechanism could be found to transfer power peacefully to a broadly representative authority.

While the resistance's fragmentation aided Najibullah, the most important reason for his survival was the regime's defence capability. The regime did not alter its counter-insurgency strategy, and continued to concentrate on defending urban areas and important lines of communication.

CHANGE IN POLITICAL FAITH

Pressure from the Soviet leadership and the political changes in Eastern Europe and the Soviet Union led the Kabul regime to transform itself. Regime functionaries and prominent leaders no longer defended socialism; they moved to a liberal, left-of-centre position on economic and social issues. Najibullah jettisoned some ideological symbols associated with the regime, changing the name of the country by dropping 'Democratic' which had preceded 'Republic of Afghanistan', and, following the pattern of party conversions in former socialist states, changing the appellation of the ruling People's Democratic Party of Afghanistan to Watan (Homeland). The winning of acceptance and credibility required the reformation of Watan, and recognition came in February 1990 when Najibullah told Parliament that the Party was not acceptable to the opposition in its old form, and suggested changes in the character and role of the Party and the constitution. The regime also discovered that Afghan nationalism and Islamic symbolism—the war-cries of the opposition—were useful tools for justifying its right to rule.

A more significant aspect of the regime's transformation was its policy shift from war against all to selective accommodation. The government targeted individuals and groups alienated by the actions of the Mujahideen and offered them concessions ranging from grants of local autonomy for larger groups to the outright distribution of weapons and money. This encouraged the tribal chiefs, some resistance commanders, and ethnic minorities in particular, to strike deals with the government. The more trustworthy were given positions in the provincial or central government.

This attempt at 'national reconciliation', as the regime called it, was actually nothing more than a return to the clientage practised for centuries by the Afghan oligarchy. However, some of the regime's initiatives, as discussed above, were successful. Local chieftains in the north joined militias organized and funded by the government, which brought them jobs, influence, and increased local power, although the loyalties of the new allies of the regime were fleeting; they successfully played one side against the other, amassing fortunes in the process.[65]

OUSTING THE MARXIST REGIME

Najibullah looked more confident in 1990, projecting his image as a unifying force while his opponents squabbled on the periphery. The population of Afghanistan, if not entirely won over by Najibullah, appeared to be tolerant of his rule, at least in the major urban areas, as long as he supplied food and provided credible security.[66] A general war-weariness in the countryside and the inability of the resistance to unite or make significant military gains contributed to his confidence, which was perhaps the single reason for the Soviets not ditching Najibullah even after their withdrawal.[67] However, Najibullah's worries never ended. Disunity within his party, several *coup* attempts from within the armed forces[68] and militant resistance continued to threaten the stability of his regime. A general condition of siege and civil war persisted, requiring the regime to rule under a state of emergency.[69] Worsening economic conditions in 1991, and finally the disintegration of the Soviet Union, which had nourished and defended the Marxist regime for more than a decade, led to its demise in April 1992.[70] Let us examine some of these factors in detail.

The most important and well-organized opposition groups remained outside the 'national reconciliation' process, adamant in their intention to unseat Najibullah. They were heartened by developments inside and outside the Soviet Union as the central government collapsed. The most important internal change, from the viewpoint of the Afghan resistance, was the shift in political power to the reformist and democratic leadership of Russian President Boris Yeltsin. Yeltsin and his close associates had been critical of Moscow's policy on Afghanistan. With the eclipse of the KGB in domestic and international affairs and the ouster of communist hard-liners from the Soviet power structure after the failed August 1991 coup, Kabul lost almost all its supporters in the Soviet Union. The regime's political and diplomatic isolation further deepened.

Even before the break-up of the Union, Moscow had changed its position on Afghanistan. In September 1991 the Soviet government announced that it would cut off military supplies to the Afghan government after 1 January 1992.[71] It also committed itself to withdrawing all military advisers from Afghanistan and to reducing economic assistance, including food and fuels. In November 1991, Moscow for the first time invited a group of Afghan Mujahideen representatives to negotiate a political settlement, and the accords that were reached formally recognized the Mujahideen as a legitimate party. To Kabul's further embarrassment, the Soviet leaders and the Mujahideen issued a joint statement calling for the 'transfer of the entire state power in Afghanistan to an interim Islamic government'.[72] As a follow-up to the meeting, the Russian Republic's Vice President, Aleksandr Rutskoi, visited Iran, Pakistan and Afghanistan in December. While his primary objective appeared to be the release of Russian prisoners of war still held by the Mujahideen, he reaffirmed Moscow's decision not to supply military equipment to Najibullah. However, Russian leaders endorsed a cease-fire, intra-Afghan dialogue, and the formation of a joint coalition government, preferably made up of moderates and non-communist members of the Afghan government.[73]

Although the former Soviet republics abstained from direct involvement in dismantling the political edifice in Kabul, denying the Afghan government military hardware, fuel, food, and political support precipitated its downfall. This was not a well-designed policy to strangle the Kabul regime, but the members of the new

Commonwealth of Independent States faced serious shortages of food and fuel themselves, and had little to offer Kabul. The Afghan government had already acknowledged that it had not received the amount of economic assistance it had been promised by Moscow.[74]

The assistance was important: economic collapse threatened Kabul more than the Mujahideen. Dependence on foreign sources—especially the Soviet Union—and total mismanagement had devastated the economy, but the government itself was the single most important factor in ruining the economy. The twelve years of fighting had caused widespread damage to the rural infrastructure. Confiscation of lands by the regime in the early 1980s had disrupted the land tenure system, and the government was unable to restore the system since it had no control over the countryside. Shortages and astronomical prices of essential goods had created economic anarchy. Gasoline, heating fuel and food were difficult to find even at the higher prices.[75]

The government claimed that in the first six months of 1991 it had achieved some of the targets set by its annual development plan; even if these claims were true, production in mining and manufacturing had declined compared to 1978 and 1979.[76] The government cited shortages of raw materials and disruption of power supplies—often damaged by resistance groups—as the causes of the slide in productivity, but the inherent flaws in centralized economic planning and the disproportionate allocation of resources to defence should not be ignored.

The country's export base has been limited to natural gas, fruits, animal skins and wool products, which together account for nearly 95 per cent of the foreign-exchange earnings. For 1990 and 1991, imports were estimated at approximately $ 1 billion, leaving a trade deficit of about $ 585 million.[77] Most of this was due to imports from the Soviet Union, which had become the country's largest trading partner.

Soviet officials in November 1991 drove a hard bargain with Kabul, insisting that it would have to meet bills for Soviet imports with shipments of its natural gas and that credit terms would be more stringent in the future[78], and in 1991 they were pressing Kabul to repay the debts incurred earlier. The end of Soviet patronage and worsening economic conditions weakened Najibullahs's already beleaguered regime, making it even more vulnerable.

The September 1991 agreement between the United States and the Soviet Union had called for a stop to the supply of military hardware to their respective clients in Afghanistan, and Moscow's decision to wash its hands of Afghanistan in general prompted the Mujahideen leaders to launch major attacks. Since February 1991, the Mujahideen had continually fought government forces in small but costly battles. With the exception of the Hazarajat in central Afghanistan, almost all regions of the country witnessed fierce fighting. In the Spring of 1991 the resistance scored a major success when it captured Khost in eastern Afghanistan—the town had been under siege for years, and the government had made it a point of pride to retain control.[79]

After this victory the Mujahideen increased their military pressure on the area around Gardez. Success here put the resistance groups in a better strategic position. Since the 1989 Jalalabad offensive, they had maintained control of seventy miles of the highway to Torkham, a border town in Pakistan, effectively denying government forces use of the Khyber Pass. The Mujahideen also compelled Kabul's forces to deploy along too many fronts and to constantly engage either in defending their turf or in launching offensives to retake lost areas. Although the regime held out against the resistance's military pressure, it could not escape disintegration under the cumulative weight of economic chaos and the end of foreign supplies of food and weapons.

Another factor that brought the regime down was the factional rift in the Watan party. Anxiety and uncertainty led individual leaders and groups from the factions to realign along ethnic and tribal lines. It started in January when Najibullah attempted to replace Tajik and Uzbek commanders in the northern provinces with more trusted Pushtuns. At this time Abdul Rashid Dostum, an Uzbek militia commander who was propped up by Najibullah, reactivated his back channels with Ahmad Shah Masud, a Tajik guerrilla commander. Some of the Tajik and Uzbek army generals, such as Mohammad Nabi Azimi, a Tajik, apparently tried to win Dostum back, but as the collapse of the regime accelerated, they began secretly supporting the formation of an alliance among the rebellious co-ethnic militia commanders and guerilla leaders in the north. This process gained impetus from Najibullah's announcement on 18 March that he would step down once an understanding was reached through the United Nations for the establishment of an interim government. He pledged to transfer all powers and executive authority to the interim government from the first day of the transition period.[80]

It is important to note that the United Nations peace plan that had been negotiated over the previous two years had received the support of all the regional powers and almost a majority of the Mujahideen parties. The plan outlined five points to serve as a basis for a settlement which would come into effect in May 1992. These were:

1. The necessity of preserving the sovereignty, territorial integrity, political independence and non-aligned and Islamic character of Afghanistan.
2. The recognition of the right of the Afghan people to determine their own future free from outside intervention.
3. The need for a credible and impartial transitional mechanism to be worked out and agreed upon through intra-Afghan dialogue. Political arrangements in the transition period would lead to a broad-based government and an end to hostilities.
4. A halt to arms supplies to all Afghan sides, by all.
5. Reconstruction of the war-ravaged country and rehabilitation of Afghan refugees.[81]

In seeking an agreement Benon Sevan, the UN Secretary-General's personal representative for Afghanistan, held several rounds of talks with all the actors directly involved in the conflict, and with their outside supporters. Owing to the internal military stalemate and changes in the direction of US and Russian policies, the two no longer being global foes, the UN diplomatic initiative, on the whole, evoked a positive response. The shaping of a broad-based interim government that would replace Najibullah was the main concern. Under the UN scheme, 150 Afghans drawn from across the political and religious spectrum were to be invited to meet in Vienna in April to start a two-stage peace process. In the first meeting they were supposed to select a group of thirty to thirty-five to contact fellow country-men both in and outside Afghanistan with a view to convening a grand assembly to name an interim government.[82]

As Najibullah started loosing control of the northern provinces, and the struggle for power began to take ethnic shape, Benon Sevan, the UN Special Envoy, with the consent of the Afghan President, on 2 April announced an accelerated settlement plan. He proposed to convene a grand assembly of up to 500 Afghan representatives on an urgent basis towards the end of the month to elect the interim government. He referred to rising fears of ethnic confrontation in Kabul as the reason for his fast-track peace initiative.[83] The plan, though not fully revealed, seemed to contain three important features; first, pre-transition arrangements which could take effect even before the convening of the Afghan assembly; second, the possibility of UN Security Council guarantees to prevent army, ethnic or ideological groups from capturing power in Kabul during the peace talks; third, that the transitional authority would be all-powerful (this would depend on the support that the plan received from the Mujahideen as they became convinced that Najibullah was ready to quit power.)[84]

Within days of this announcement, the rebel militia commander Dostum began setting up an 'autonomous' administration in Mazar-i-Sharif by forging an alliance with Ahmad Shah Masud.[85] As defections deepened the crisis of government, the non-Pushtun army generals and Party leaders forced Najibullah, a Pushtun, to resign on 16 April, before the United Nation peace plan could be implemented. Had the plan as envisaged by the UN negotiator been implemented it would not only have facilitated a smooth, peaceful and orderly transition but would also have paved the way towards a stable political settlement. The military *putsch* in Kabul, and the advance of the joint forces of Masud and Dostum from the north capturing strategic towns and military bases in the wake of hectic negotiations by the UN representatives, has raised a number of questions. Why did Masud enter into alliance with Dostum and the army generals at a time when arrangements to replace Najibullah through the UN plan looked credible? How important was the ethnic factor in determing the course of this alliance?

It was no coincidence that all the key figures who were instrumental in removing Najibullah were non-Pushtuns.[86] The Jihad

Council which Masud set up had no Pushtun either.[87] There is no doubt that though Masud preferred to camp outside Kabul with his force, while insisting that the Mujahideen parties based in Peshawar, Pakistan, form an interim government to fill the vacuum, the forces of his ally Dostum had already taken over control of the airport and strategic points in the city. As events unfolded, three things emerged increasingly clearly. First, Masud and his coalition partners saw little gain for themselves in the UN-sponsored political arrangements, as power would have been more diversified. Second, they wanted to outflank Hikmatyar, who they feared would have struck a deal with the Pushtun generals and moved into Kabul with their support. Third, it appeared to be a historic opportunity for the minority Tajik, Uzbek, Hazara and Ismaeli leaders to establish a new balance of power with the majority Pushtuns. Finally, there emerged a salience of power politics over ideological issues after the collapse of the Marxist regime.

NOTES

1 *Times of India,* 11 February 1981.
2 *Tass,* from Kabul, 29 December 1979, in FBIS/SU, 31 December 1979, pp. Dl-5.
3 Yuri Gankovsky (ed.), *A History of Afghanistan* (Moscow, Progressive Publisher, 1985).
4 Karmal's press conference reported by Radio Kabul, 7 January 1980.
5 Radio Kabul, 5-6 January 1980.
6 The Soviets continued to express concern over polarization. *Muslim,* 9 August 1987.
7 *See,* report of the Central Committee Plenum held in October 1989. *Kabul Times,* 26 October 1989.
8 Radio Kabul, 22 September 1984, in FBIS/SA, 4 October 1984, pp. C1-8.
9 *Kabul New Times,* 12 August 1985.
10 *Afghanistan Forum,* November 1985, p. 5.
11 This is a rough estimate as no authoritative source on the strength of the leftist movement at the time of the revolution exists.
12 Author's interviews with Afghan resistance leaders in Peshawar on 5 January 1990.
13 Author's interview with Professor Yuri Gankovsky on 22 December 1989, in Islamabad.
14 Radio Kabul, 20 January 1986, in FBIS/SA, 24 January 1986, p. C1.

15 Radio Kabul, 20 January 1986, in FBIS/SA, 24 January 1986, p. C1.
16 Radio Kabul, 3 February 1986, in FBIS/SU, 4 February 1986, pp. D3-4.
17 Radio Kabul, 25 April 1985, in FBIS/SA, 1 May 1985.
18 Bakhtar, 14 September 1985, in FBIS/SA, 16 Sept. 1985, p. C1.
19 About 300 delegates were from Pakistan. Henry Bradsher, 'Stagnation and Change in Afghanistan,' *Journal of South Asian and Middle Eastern Studies,* Vol. 10, No. 1, Fall 1986, p. 23.
20 Radio Kabul, 16 September 1985, in FBIS/SA, 20 Sept. 1985, p. C2.
21 *Muslim,* 14 April 1986.
22 Author's interviews with officials of the Government of Pakistan in January 1990, in Islamabad.
23 *Kabul Times,* 13 November 1989.
24 Radio Kabul, 11 January 1986, in FBIS/SA, 14 January 1986, p. C1.
25 The author's discussion with Afghan dissidents in Peshawar, 4 January 1990.
26 *Muslim,* 1 January 1990.
27 KHAD agents killed 120 Mujahideen in Fariab province in May 1982. *Afghan Realities,* No. 6, July 1982.
28 Author's interviews with Mujahideen representatives in Peshawar, on 2 January 1990.
29 *New York Times Magazine,* 24 March 1985, p. 53.
30 In an interview with this author, an official of the Media Information Department of the Afghan Interim Government claimed that about 5000 Afghan children were sent to the Soviet Union. This is certainly an exaggrated figure. Interview was held in Peshawar on 3 January 1990. The Mujahideen government later indicated the number as 4,000. *Muslim,* 15 May 1992.
31 Ibid.
32 From Kabul University faculty alone 36 were executed, 6 imprisoned and 276 left the country. *See,* Sayed Mohammad Yusuf Elmi, 'The Impact of the Sovietization on Afghan Education and Culture,' S B Majrooh and S M Y Elmi, (eds.), *Sovietization of Afghanistan,* op. cit., p. 80.
33 For this information I have benefited from Professor S M Y Elmi, who is a former faculty member of Kabul University. Ibid., pp. 73-9.
34 Jean J Kirkpatrick's statement in the U N General Assembly on 24 November 1982. *Department of State Bulletin,* 24 November 1982.
35 *Far Eastern Economic Review,* 13 July 1989, p. 17.
36 Edward Girardet, 'Soviets Grab Afghan Resources, Saving their Own,' *The Christian Science Monitor,* 22 December 1982.
37 Comments of former Afghan officials of Mines and Industry Ministry, in *Muslim,* 21 March 1987.
38 *Kabul Times,* 16 December 1989.
39 Ibid.
40 *Pravda* (editorial), 21 December 1985, in *CDSP,* Vol. 37, No. 541, 15 January 1986, pp. 18-19.

41 *Far Eastern Economic Review,* 12 December 1985, p. 30.
42 Khalid Mehmud, 'The Afghan Conflict: New Soviet Overture,' *Muslim,* 29 December 1985.
43 *Muslim,* 28 April 1986.
44 Moscow Television, 25 February 1986 in FBIS/SU, 26 February 1986, p. 31.
45 'How all this happed: Expressions of Najibullah,' *Kabul Times,* 18 December 1989.
46 *Izvestia,* 13 March 1986.
47 Najibullah is an Ahmadzai from Paktia province. His grandfather was a chieftain. *Current Biography,* June 1988, pp. 39-41.
48 *Muslim,* 5 May 1986.
49 Kabul Radio, 1 May 1986, in FBIS/SA, 2 May 1986, p. C1.
50 *Muslim,* 7 May 1986.
51 Anahita Ratebzad organized women's demonstrations. *New York Times,* 12 May 1986, p. 47.
52 Rahimullah Yusufzai, 'Dr Najibullah—the Man,' *The Muslim Magazine,* 17 June 1988.
53 *See,* for instance Najibullah's interview with Swedish correspondent Pierre Allison in *Kabul Times,* 19 November 1989.
54 *See,* for instance, *Kabul Times,* 17 December 989.
55 In an interview Najibullah told Tass that his government was paying salaries to 11,570 mullahs in 5,167 mosques. *New Times,* 10 November 1986, pp. 12-13; 3 billion Afghanis were spent for religious purposes. *Kabul Times,* 14 December 1984.
56 *Muslim,* 20 December 1989.
57 *See,* Najibullah's interview with Pakistani journalist Ayaz Amir, *Herald,* February 1987, pp. 31-3.
58 *Keyhan,* 19 April 1990, p. 2 in FBIS, 7 May 1990, p. 45.
59 *News,* 8 June 1991.
60 *New York Times,* 26 February 1990, p. 9.
61 Radio Kabul, 22 May 1990, in FBIS, 23 May 1990, p. 32.
62 *Muslim,* 17 February 1991.
63 *Muslim,* 27 November 1990.
64 *Muslim,* 26 November 1990.
65 Gilles Dorronsoro, 'The Militia in Afghanistan,' *Central Asian Survey* Vol. 8, No. 4, Fall 1989, pp. 95-108.
66 James Rupert, 'In Kabul a Shift of Opinion,' *International Herald Tribune,* 28 July 1989.
67 Author's interview with Pacha Gul Wafadar in Peshawar, 2 January 1990.
68 The Afghan secret police arrested 127 suspects from the armed forces in December 1989. *Muslim,* 1 January 1990.
69 *Kabul Times,* 22 November 1989.
70 *New York Times,* 18 April 1992, p. 1.
71 *Los Angeles Times,* 14 September 1991.

72 *Pakistan Times* (Rawalpindi), 17 November 1991.

73 *Wall Street Journal,* 4 September 1991.

74 Anwar Dost, the Food Minister of Kabul, stated that his country received only one third of the wheat the Soviet Union had promised. *Afghan News* (Peshawar), 15 September 1991.

75 Edward A Gargan, 'Kabul Journal: Winter, Too, Is Hell; Afghans Teeter on Abyss,' *New York Times,* 20 March 1992.

76 Government of Afghanistan, *Afghanistan 1990,* report prepared for United Nations Conference of the Least Developed Countries (1990), p. 7.

77 Ibid, p. 23.

78 Hong Kong, AFP in English , 8 November 1991 in FBIS, 8 November 1991, p. 42.

79 *News* (Rawalpindi), 2 April 1991.

80 *New York Times,* 19 March 1992, p. 3.

81 *Muslim,* 22 May 1991.

82 Edward A Gargan, 'Afghan President Agrees to Step Down,' *New York Times,* 19 March 1992, p. 3.

83 Steve Coll, 'Fresh Hope Raised for Afghan Peace,' *Washington Post,* 3 April 1992.

84 According to the media reports, even Gulbadin Hikmatyar, a hard-line guerrilla leader, supported this plan. Benon Sevan was optimistic that peace was near in Afghanistan in April. He vowed to make sure that 'no one—I repeat, no one—will dare to destroy the peace process,' See the report by Steve Coll, *Washington Post,* 3 April 1992.

85 Barbara Crossette, 'Defections Deepening Kabul's Isolation,' *New York Times,* 9 April 1992, p. 3.

86 There were: Generals Abdul Momin, Mohammad Nabi Azimi, Mansur Naderi and Abdul Rashid Dostum. Among the Watan Party leaders, Farid Mazdak and Abdul Wakeel played key roles.

87 Ahmad Rashid, 'Mujahideen government may take several days to take off,' *Nation,* 21 April 1992.

Chapter 7

The Afghan Resistance

The character of the Afghan resistance is among the most misunderstood phenomena of insurgencies. Its portrayal has ranged from that of a tribal uprising against 'progressive socio-economic reforms' to a movement for liberating Afghanistan from Soviet domination and communist rule. Assessments of the Afghan resistance have frequently centred on its lack of unity, organization and national leadership. While some critics have argued that the Afghan resistance never developed into a true national liberation movement, others have projected rather uncharitable images. Personal rivalries, factionalism, infighting and ideological conflicts have been blown up out of proportion. While some of these observations might be approximately objective and true, they neither explain the whole truth nor reflect the complexities inherent in the structures of Afghan society. Undoubtedly, the Afghan resistance was never united during its entire period of active armed struggle, but for more than nine years it denied the Soviet occupation forces effective control over the country. This is not to downgrade the significance of a unified political and guerrilla structure which would, in any case, be crucial to the final outcome of the conflict, but it is necessary to remind ourselves that, though fragmented, the Afghan resistance succeeded in achieving one of the central aims of any guerrilla war—military stalemate. It was the Mujahideen factor that turned the Soviet military intervention into its 'bleeding wound.' When the occupation war became costly, the rhetoric of 'international duty' gave way to a realistic assessment of Moscow's military involvement in Afghanistan.

One is always tempted to evaluate political and military events from a comparative perspective, but such an attempt can lead to erroneous conclusions if the peculiarities of a society are not taken into account. Thus, compared to the Algerian resistance or the Vietcong the Afghan resistance would appear to lack the essential ingredients of a war of national liberation. The fact is that while the goals of resistance movements might be similar or even identical, they take different political forms because they take shape in different social and political milieux. Geography, the technology of counter-insurgency warfare and the power projection capabilities of the interventionist state are the factors that need to be considered for understanding differences among the organizational patterns of the resistance movements and their strategies. Depending on the nature of the actors involved in the conflict and the stakes of the external powers, the prevailing regional and international setting have differing impacts on the resistance movements. Thus, given the domestic origins of the conflict, it would be more enlightening if the Afghan resistance could be understood in its own social and historical context.

POLITICAL AND SOCIAL ORIGINS

The conflict between the communist government and the Mujahideen resistance was as much a manifestation of political polarization in the country as a struggle for liberating Afghanistan from the Soviet occupation. The dominant groups that led the Afghan resistance represented an old ideological conflict which took shape during the constitutional period (1964-73). The introduction of the electoral process, Parliament and the relative freedom of the Press marked the beginning of a rift between the Islamist and Marxist groups. One can trace the begining of this rift to the disagreement on the restructuring of the Afghan political order between the modernists and the Islamic-minded intellectuals. Old debates and controversies strongly influenced the organizational pattern of various political groups. The politics of the Marxist groups has been explained in detail in Chapter 2. The Islamist elements, alarmed by the views of Khalq and Parcham, reacted sharply by urging the monarchy to ban Marxist organizations. The traditional *ulema,* benefiting from the relatively liberal political atmosphere in the country, succeeded in capturing a far greater number of seats in Parliament than their Marxist foes did, and so became more vocal and assertive in parliamentary debates. The traditional Islamic elements, though they did not develop any significant organizational structure, were able to effectively demonstrate their potential political power by inspiring agitation in Kabul against the activities of the Marxists in 1969.

Islamist intellectuals, scholars and student activists—largely from the government educational institutions—began organizing a broader Islamic revolutionary movement rather than merely responding to the politics of Marxist groups. Nevertheless, Soviet influence with the Afghan communist movement provided an immediate context for the Islamists to develop their movement on modern lines. Most of the prominent leaders of the Islamist resistance groups on the scene today were founders of the Islamic movement in Afghanistan. Rabbani, Sayyaf, Khalis, Hikmatyar, and commanders Masud and Ismael worked together in the initial phase. Even though Islam as a political ideology was a new movement, it caught the imagination of large numbers of educated Afghans. The leaders of the Islamic movement were charismatic and dedicated individuals, but Islamic symbols also worked effectively in Afghan society. Within a relatively short period of time, students affiliated to the Islamist movements took over control of Kabul University's Student Union, which had for a long time been functioning almost as a branch of the Marxist faction led by Babrak Karmal and Anahita Ratebzad. From 1967 onward, all the educational institutions witnessed the ideological rivalry between the Marxist and Islamist youth, often leading to bloody battles.

The Islamists proceeded to establish a nationwide organizational network by approaching some of the traditional *ulema* and Islamic-minded influential figures in the provinces. At the same time, they devoted their resources to the preparation of literature and propaganda material. It was partly the growing influence of the Islamist group and the insistence of the traditional madrassah (Islamic religious school)-based *ulema* on the complete enforcement of *sharia* (Islamic law) that pushed the modernists, nationalists and *Parchami* socialists into forming an alliance in 1973.

This informal alliance led by Sardar Daud overthrew the monarchy and declared Afghanistan a republic. The Islamists, not the old oligarchical monarchist groups, became the main target of the Daud regime. Hundreds of Islamist activists were arrested, and some summarily executed. A small number of them, led by Hikmatyar, Masud and Rabbani, escaped to Pakistan, from where they launched their guerrilla movement in 1975. Zulfiqar Ali Bhutto's government patronized the Islamist guerrillas more for political than ideological reasons. His government provided funds and equipment and imparted training to the exiles, who would later emerge as very important and successful commanders in the liberation war against the Soviet forces.

It is important here to examine why the secularist and apparently socialist Bhutto should aid Islamist guerrillas who he knew had links with his staunch opponents in the Jamat-i-Islami of Pakistan. First, Daud had reopened the issue of Pushtunistan and the Durand Line, which Pakistan has long considered as settled. However, he opted to play the Pushtunistan card more to establish his nationalist credentials and to generate popular support for his regime than for any other tangible effect on the common boundary. Second, the Afghan government was involved in aiding the largely Baloch and Pushtun National Awami Party (NAP), which Bhutto had managed to oust from power. The NAP had the majority of seats in Balochistan, a province which is contiguous with Afghanistan. As a result of the dissolution of the NAP government in Balochistan, the North-West Frontier Province (NWFP) government, which was a coalition of NAP and Jamiat-i-Ulema-i-Islam, resigned in protest. The political confrontation between Bhutto and the NAP took a tragic turn with sabotage and bomb explosions in various parts of the country. A simple law and order situation deteriorated quickly into a full-fledged insurgency in Balochistan.[1] All the prominent leaders of the NAP were jailed, and trials were instituted against them on the grounds of 'treason.' The Pakistan Army was deployed in Balochistan to fight a costly and destructive counter-insurgency war. The Bhutto government suspected, and now it is a well-known fact, that his opposition in the NWFP and Balochistan was supported by Daud.

For no other reason than to bring Daud to a compromise did Bhutto support the Islamic militants of Afghanistan.

The first armed encounters of the Islamists with the Daud regime took place in 1975. While Ahmad Shah Masud launched a campaign in Panjsher, Hikmatyar acted simultaneously in Paktia. The objective was to stir up an uprising against the Daud regime. As it transpired later, the guerrilla leaders had neither the organization nor a sufficient number of fighters to topple the government. They had had very little time between their self-exile in Pakistan in 1973 and their sporadic attacks in 1975 to establish linkages with the local population, who at that point were not willing to take up arms against the Daud regime. The attempt failed. Relations between Afghanistan and Pakistan improved by 1976 and Bhutto withdrew his support for the Islamists. In the aftermath they lay low, focusing on self-education in guerrilla warfare and devoting much attention to organizing and training disciplined party workers. Two points emerge from these developments. First, the Islamist resistance groups arose out of the circumstances made possible by the relative liberalization of the constitutional period. Second, the ideological antagonism of the Islamists towards the Moscow-backed communists was as old as the first initiatives by the Soviet Union to penetrate the Afghan state in the 1950s.

The Islamists were the first to point to the dangers of growing Soviet influence in Afghanistan. They had a better historical perspective on Soviet moves than the traditional *ulema,* who preferred to remain either apolitical or tied to the oligarchical power structure. The events of the April 1978 *coup,* and the Soviet military intervention in 1979, emerged as vindication of their apprehensions. Nevertheless, the Islamists did not emerge as a strong political force in the 1970s. Several factors inhibited their growth. First, the founders of the movement preferred a rather small but indoctrinated and disciplined cadre to a mass organization. Secondly, a large number of Islamist workers and leaders were executed after the Islamist guerrilla attacks in 1975.[2] Third, the liberal period was too short to give the Islamic movement a popular basis. Finally, the Islamists faced opposition from the traditional *ulema,* which had previously monopolized all conventional arenas of religious politics.

The Islamist group remained small and politically ineffective until Afghan society experienced the upheaval following the Saur revolution in 1978. Reforms introduced by the avowedly Marxist regime ignited widespread revolt. Self-characterization by the regime as 'proletarian and founded on the basis of the epoch-making ideology of the working classes'[3] further provoked popular sentiment against it. Apparently impressed and inspired by the socialist restructuring in the Soviet Union, the Afghan Marxists embarked upon transforming Afghan society from what they abhorred as hierarchical, tribal and feudal into a modern egalitarian one. The Afghan leftists, like similar movements in other Third World states, regarded the traditional power structures as obstacles to the revolutionary change in their own particularistic image. Mesmerized by the cliches, the Afghan Marxists blindly set their faith in the 'revolutionary potential of the masses,' assuming that the positive value of reforms would generate popular support and legitimacy for the regime. They neither made an objective analysis of the social and political forces in Afghan society nor expended any meaningful effort on intelligently educating the masses on the necessity of reforms. What explains their feckless idealism is the fact that the Afghan leftist movement was primarily urban-based and totally cut off from the realities of an overwhelmingly rural, agrarian culture. The messianic and mechanistic approach to revolution and reforms merely exposed the perceptual gap that existed between the Marxist leaders and the Afghan masses. Their idiom and reference group were so alien that political communication with the population became impossible.

The domestic policies and measures largely focused on destroying the traditional structures of power and influence. This brought the Marxist regime of Taraki and Amin into confrontation with the privileged social classes, such as the tribal chiefs, the land-owing aristocracy and the religious establishment. Fearing a backlash from the traditional elite, the regime acted hastily to produce positive results. In their zeal for a new Afghan society, the Khalq leaders committed two fundamental mistakes. First, the very concept of socialization and forcible nationalization of private property was incompatible with the Afghan political culture. Second, they underestimated the power and influence of the traditional elite. The Afghan individual who participated in various overlapping social networks regarded his allegiance to the norms of the well-structured system of client-patronage as a set of moral, religious and social obligations. The population that was supposed to benefit from the reforms did not see them as socially appropriate because they lacked the sanction of custom and religion, which had a strong influence on the personal values of ordinary Afghans.

As the reform measures provoked fierce opposition, the regime's dogmatic functionaries used indiscriminate violence in the countryside. In Afghanistan, individuals and families function within the much larger framework of kinship and tribe. The Khalq regime never had a calculated strategy for isolating the traditional elite from their immediate social networks. Persecution of opponents by the state security forces, or seizure of property from a member of a particular social network, provoked a more or less collective response. The age-old tradition of *badal* (revenge) resulted in a spiral of repression and resistance. As a result, the Afghan society lost its traditional equilibrium. The balancing elements may never be the same, even if political stability is restored to the country.

The opponents of the regime, mostly from the traditional power structure, skillfully capitalized upon the mass resentment against the Khalq regime. As in the past, they invoked Islam and Afghan culture to rally support against the regime. The atheistic and anti-Islamic image of the Khalq regime volatilized the countryside. As the authority of the government became tenuous in the rural areas, localized resistance groups began to emerge along ethnic, tribal and religious lines.

Identification of the Khalq regime with the Soviet Union was another factor which fuelled the mass uprising. The Khalq leaders rhetorically and quite deliberately linked their revolution to Bolshevism. They servilely toed the Soviet line on international issues. More than the genuineness of its revolution, this dependent Marxist group capturing power in a neighbouring country was what provoked the sympathy and support of the Soviet Union. A flurry of agreements enabled the Khalq regime to receive Soviet economic and military assistance which the *Khalqis* needed to stabilize their control. Soviet advisers in large numbers followed the aid agreements. The Soviet influence over the Khalq regime was deep and visible. This lent credibility to the conspiratorial view of the Saur revolution projected by the opposition. The Marxist symbolism of the regime and its heavy dependence on the Soviet Union provided effective propaganda tools for the opposition in mobilizing the masses against the regime.

After the killing of Taraki, Amin, known for his ideological hard line and ruthlessness, reacted to the national uprising in extreme confusion. He relied more heavily on repression and the extensive use of force than on winning the support of the masses and influential groups through political means. The propaganda style of his regime erected strong barriers against establishing its Afghan credentials.[4] There were considerable doubts about the ability of the Amin regime to restore order, much less to survive against the rising power of the resistance. Threatened by the Mujahideen, alienated from the people and in conflict with Parcham, Amin's only power base remained the top leadership of the Afghan army. Nor did the Soviets trust him after his execution of Taraki. Since the pro-Moscow Parcham was weak and scattered, the Soviets took it upon themselves to remove him.

The Soviet intervention, as we have explained in Chapter 3, altered the entire political and military scene of Afghanistan. The worst fears that the Islamist opposition to Daud had expressed much earlier proved true with the Soviet occupation of the country. A new phase in the Afghan resistance movement began with the Soviet military presence and its active participation in the war. The character of the resistance became truly national as civil and military personnel of the regime, more nationalistic PDPA members and even secular intelligentsia reacted sharply to the Soviet invasion. In Afghan history nothing has affronted the honour of the nation more than the presence of foreign forces on its territory. The ethnic, linguistic and tribal divisions characteristic of Afghan society have never militated against a national sentiment in time of crisis, such as foreign invasion, and owing to these divisions, hundreds of resistance groups emerged, without co-ordinating guerrilla strategies. The influential *qabila* (tribe) Sufi (mystic) and *ulema* networks made a unanimous call for

Jihad (holy war or struggle) against the Karmal-Soviet forces and organized fighters in their respective areas of influence. Such was the state of frenzy that even without any leadership people took up arms to fight and die rather than see their country violated by the Soviet invaders.

ISLAM AND THE RESISTANCE

Almost all the resistance groups invoked Islam as a source of inspiration for waging war against the Soviet Union and its Kabul proteges. It was not the first time that the Afghan Muslims had turned to religion for motivation. Islamic history is replete with similar instances when Muslims have legitimized their wars against aggressors by religious sanction. Even dynastic and territorially-motivated battles against fellow-believers have been given religious justification. The reason is that an appeal to Islam gives the Islamist fighters a greater sense of direction. The spirit of martyrdom in the broad Islamic view of life and death induces tremendous motivational effects in the individual. Both in the defensive and in the offensive wars of the Muslims, religion has never been separate from the process of conflict.

The mobilizing effects of the religious appeal in the wars of resistance have varied according to the historical context of the occasion and the degree of religiosity of the civil society concerned. Afghanistan is one of the few Muslim countries where the practising of Islam is somewhat customary. Therefore, the religious orientation of almost all the resistance groups was a natural reflection of the fact that Islam has been a powerful cultural symbol in Afghan society. It was also pragmatic and sensible for the Afghan resistance to express national liberation mythology in a religious medium. Therefore, it is necessary that we refer to the entire resistance as Islamic at one level of explanation. But it would be simplistic and obscure the facts if a clear distinction were not made between traditional and revolutionary Islam in the political context of Afghanistan. Major divisions within the Afghan resistance reflected the differences between traditional and revolutionary views of Islam.

Islam, though widely practiced at the individual level, never penetrated the Afghan state structure in the ideological sense. Ever since the establishment of the Afghan state in 1747, Afghan rulers have acquired and maintained power largely through the manipulation of tribal alignments. They seldom sought political legitimacy through the Islamic factors alone, but they respected the religious establishment more as a religious and cultural symbol than in any quest for the justification of their authority. Afghan kings, with few exceptions, patronized prominent *ulema,* sought their advice, appointed them in positions of *qadi* (religious judge) and made big contributions to the religious *waqfs* (endowments). In the Afghan tradition, the state and religion remained two separate and mutually exclusive domains.

The political character of Islam in Afghan society was shaped by two factors. First, although the Afghan political system never established a strong state, religion remained separate and subordinate to the political interests of the rulers. Without being visibly antagonistic, the secular authority skillfully managed the religious networks through a dual strategy of autonomy and selective co-option. Second, the traditional religious establishment of Afghanistan restricted itself to the spiritual training of the individual, seeking to inculcate conformation of social behaviour to Islamic values. While traditional Islam remained concerned with the profession of faith, the Durrani aristocracy which controlled the Afghan state until 1978 managed its affairs through tribal law and custom[5] with a smattering of Islam. The Afghan tribal code, though not strictly *sharia* (Islamic law), has evolved within the civilizational context of Islam, and in essential matters is more or less in agreement with its religious spirit.

Broadly speaking, three traditional institutions which preceded the formation of the Afghan state have exercised religious influence over Afghan society. These institutions are the Sufi, *ulema* and Sayyid (descendants of the Prophet).[6] Islamic beliefs, practices and rituals and the values of ordinary Afghans have been shaped by the powerful personalities of the leaders of these traditional institutions. The boundaries of their ethnic and geographical constituencies have generally overlapped, thus reinforcing the traditional Islamic outlook. Social esteem and the degrees of their influence vary widely according to the nature of the institution and personal standing of the leaders.

The Sufi institution has generally commanded a cumulatively greater mass following.[7] The Sufi orders as they have evolved are no longer restricted to the religious sphere. The families of the founders of the Sufi orders have exercised considerable social and political influence over their followers. The Sufi brotherhoods have served as a unifying force among adherents belonging to diverse social, ethnic and geographical groups. By virtue of cross-cutting influence, high social esteem and the financial powers of the shrines, the leaders of the Chishtia, Nakshbandia and Qadriyya Sufi orders[8] have been prominent national figures in Afghan history. Some of them were linked to the Afghan ruling class through matrimonial ties. The Afghan state structure unfailingly showed reverence and recognized their importance by distributing financial favours. Two of the Afghan parties are headed by leaders of Sufi orders: Sibghatullah Mujaddedi and Sayyid Ahmad Gailani belong respectively to the Naqshbandia and Qadriyya orders.

The Sayyids, who claim to be the descendants of the Prophet of Islam, enjoy tremendous personal prestige in Afghan society although the political influence of the Sayyid families has overall been limited. The following of an individual Sayyid has depended much on his charisma and religious standing; his participation in other networks would considerably enhance his influence.

The *ulema,* regarded as men of Islamic learning and trained scholars of Islamic law, have been no less an important religious institution than the Sufi orders. The *ulema* need to be distinguished from the *mullahs,* who have a rudimentary knowledge of Islam, and usually fall into the category of professionals, subservient to the local authority of the Khan or village chief. As the Sufi brotherhoods operate from the shrines, the *ulema* are based at the madrassah. The traditional madrassahs are often given the title of *jamiah* (University). The core texts and curriculum of these institutions have for centuries changed little. Several madrassahs, varying in religious orientation and prestige according to the influence of their founder or head, train young scholars to undertake further expansion of the madrassah networks. The *ulema* have the powerful arena of the mosque which they use every Friday to address congregations of believers. The mosque, particularly during political turmoils, has emerged as the centre of political activities in a number of Islamic countries, but whether a mosque becomes an organizing point or not would depend on who controls it, and what are the inclinations of its *khatib* (one who addresses the congregation). The traditional *ulema* and their mosques were never politicized until the communists captured power. Also, the war of resistance has changed their roles. Most of the prominent *ulema* and their *talibs* (pupils in religious schools) spurred the national uprising against the communists and the Soviet intervention. They continue to play a major role in the political affairs of contemporary Afghanistan.

The majority of the traditional *ulema* are linked to the Sufi networks, which they regard as truly representative of Afghan Islam. Other influences, which have crept in through the Deoband tradition[9] and the Wahhabi movement,[10] have introduced rivalry among the *ulema*. The traditional *ulema* claim that these influences are alien to Afghans who, they assert, practise the religion of Sufis, not of Wahhabis. There has been so much controversy about the Wahhabi movement that even the Marxist regime attempted to benefit by branding the entire resistance as Wahhabi.[11] More for reasons of preserving their influence than because of any fundamental differences on *sharia,* the traditional religious networks have chosen to attack the legitimacy of newer Islamic influences in their society—their practices and attitudes are much closer to the popular religion which they have influenced over the centuries.

In the present century, the overall influence of the traditional religious networks declined for several reasons. With the advent of modern educational institutions, the graduates of the madrassahs found it hard to compete for placements in the expanding state bureaucracies. Secondly, the development of a state structure produced a new middle class of secular intellectuals and professionals. The modernist and western outlook of this class shifted the intellectual discourse to the ideologies of nationalism, individual rights and socialism. Finally, the new intelligentsia in Afghanistan had organic ties with the state and enjoyed considerable power and influence. The Press and other modern arenas of dialogue were effectively controlled by the state intelligentsia. Over time, modern professional and state institutions gained considerable prestige, attracting young Afghans to secular educational establishments. Equally important was the economic power of the new educated elite. Conversely, the traditional *ulema* and their madrassahs offered little in the way of social or economic mobility.

A more serious challenge to traditional Islam came from the revolutionary Islamists and their political doctrine. Unlike the traditional *ulema* they are the product of secular educational institutions, and have addressed economic, social and religious issues in political terms. More like modern intellectuals, their points of reference are the state and political action. The Islamists in Afghanistan, as elsewhere, have preferred to talk of Islam as an ideology and as a political system. In their overall scheme of thought, Islam is a *deen* (code of life), which is not only capable of governing social relations, but is also a legitimate source for guiding state policies on the modern issues of planning, economic development and the maintenance of a just order in society.[12]

Islamist thought, compared to the traditional outlook, is revolutionary in the sense that it projects a unified notion of both state and society as well as the spiritual and the material. As far as the juridical sources of Islamic ideology are concerned, they are as fundamentalist as the traditional *ulema*. The Quran, *sunat* (tradition of the Prophet) and *sharia* are undisputably the common frame of reference for all shades of Islamic opinion, but the Islamists think in terms of the general fundamental principles to be derived from these sources and a modern Islamic polity shaped in the light of the contemporary objective conditions of the Muslim societies. They have tended to raise political action to the level of a religious duty. Abul Aala Maududi, the founder of the Islamist movement in Pakistan who influenced Islamist thought in Afghanistan, has argued that the objective of Muslims must be the setting up of an Islamic state.[13]

While the Islamists use the western vocabulary of ideological discourse, they reject western thought and infuse every notion of political life with Islamic concepts, to the extent that they talk of Islamic economics and even Islamic sciences.[14] Party concepts such as cadre, indoctrination, political work and quest for political power have been equally Islamicised in content and motivation. The role of the party and political organization in Islamist thought is central to political action and the Islamic revolution. These are radical departures from the traditional Islam of Sufis and *ulema,* who paid little attention to the politics of Islam.

THE ISLAMIST PARTIES

The nucleus of the Islamist political movement developed in the Faculty of Theology at Kabul University some time in 1958. A small group of professors inspired by Dr Ghulam Muhammad Niazi met regularly in a calendistine study circle. These meetings in closed circles came to be known as Tehrik-i-Islami (Islamic movement). Formally, the group was called Jamiat-i-Islami (Islamist party). Islamic intellectuals, such as Sibghatullah Mujaddedi and Minhajuddin Gahiz also became associated with the Islamist group. Professor Burhanuddin Rabbani and Maulvi Yunis Khalis, who were graduates of Al-Azhar University (Cairo), joined the movement and became its core members.

Afghanistan's contemporary Islamist thought, with the exception of the Shiite sect, has been influenced by two Islamic movements, the Ikhwan-ul-Muslimeen (Muslim Brotherhood) of Egypt and the Jamat-i-Islami (Islamic Party) of Pakistan. The works and thought of Sayyid Qotub and Abul Aala Maududi have influenced the structure of the Islamist movement.[15] The Ikhwan and Jamat should not be construed as two competing movements; rather, not only do their political views of Islam converge but also they consider themselves as part of a pan-Islamic movement pursuing the common objective of bringing about the total transformation of Muslim society. Differences in their approach or strategy for political action are largely due to the unique political settings within which they operate. The Afghan Islamist movement has invariably been influenced by the political philosophy of the Jamat and Ikhwan. Much of this influence came through personal encounters and the study of literature. Similarly, the Afghan Islamists learned the political skills of the Jamat and Ikhwan and imitated their highly centralized style of political organization. As a first step, the Afghan Islamists translated some of the major works of Sayyid Qotub and Maududi, besides publishing their own treatises on Islam as an ideology and political system.

The political environment in which the Islamist movement operated was oppressive, and the hold of the traditional *ulema,* who happened to be the recognized spokesmen of Islam, was pervasive. Therefore, the Islamist intellectuals and scholars focused their attention on the essential preparations for widening the influence of their movement. Dialogue with like-minded *ulema,* intellectuals, and young students with Islamic inclinations figured prominently in their attempts to seek adherents. While remaining in the background, the Islamists at Kabul University encouraged the students to launch the first front organization. Sazman-i-Jawanan-i-Musulman (Muslim Youth Organization) was created to confront the leftist student groups and counter their rising influence. By 1970, the Muslim Youth Organisation had established itself as the dominant political force in the educational institutions of the capital. Ahmad Shah Masud, Habiburrehman, Gulbadin Hikmatyar, Ishaq and Nusrullah Emmat were prominent young Islamist activists whose dedication and personal qualities attracted students to join their movement.

Encouraged by their success, the young Islamists reached out to the *talibs* of the traditional madrassahs, as well as the radical *ulema* and the school teachers in the countryside. Preaching in religious forums such as mosques and distributing propaganda leaflets, along with engaging sympathetic and receptive persons in political argument, proved to be effective means of conducting political work. In outlook and activities, the young Islamists appeared to be far more radical than their mentors. This is evident from their open opposition to many traditional *ulema* for their conservatism and apolitical posture. In their speeches and propaganda material they protested the Soviet influence in Afghanistan and attacked the monarchy and its oligarchical rule. In opposition to nationalist groups, they advocated a pan-Islamic concept of the *Ummah* (Muslim commonwealth). Their modern political style and their influence in the educational institutions, however, did not help the Islamists much in penetrating traditional society. Their attitude towards the traditional religious establishment provoked fierce attacks. The traditional *ulema* questioned both the politicization of Islam and the legitimacy of its Islamist ideologues.

When the Islamist movement was still in a primary developmental stage and had not yet taken root at the popular level, Daud, with the co-operation of the army and Parcham, took over power in 1973. The coalition led by Daud was extremely hostile to the Islamist movement. Under a reign of oppression the Islamists either went underground or fled to Pakistan. The movement was crippled before it could take off in a political sense.

The unity of the Islamist movement was the first casualty of the crisis generated by Daud's persecution, which sent its leaders into exile. Opinions within the movement differed widely on political strategy. The radicals, mainly with Islamic Youth experience, wanted a military confrontation with Daud. Hikmatyar and Masud found the political climate in Pakistan favourable and went all out for organizing a guerrilla force. On the other hand, moderate and more experienced leaders such as Dr Niazi, who was imprisoned after the *coup,* and Rabbani wanted a political compromise with Daud. Rabbani even offered co-operation with Daud on two conditions: a) release of all political prisoners, and b) expulsion of the *Parchamis* from the government.[16] Attempts to seek reconciliation failed. It was perhaps a pragmatic move on Rabbani's part to open a dialogue with the government seeking the release of about 200 Islamist prisoners who were a political asset, having been trained over a period of a decade. Later, after the Saur revolution, all of them were executed by the *Khalqis*.

A formal split in the Islamist movement occurred in Pakistan in 1975. Two separate groups began to emerge, one around Rabbani and the other around Hikmatyar. Complex factors of ethnicity, social background, leadership ambitions and strategy divided a once unified group welded together by ideological ties. The lessons of the failure of the 1975 uprising were never lost on Rabbani. In his view a broader alliance or a sort of united front had to be worked out among the Islamist intellectuals, traditional *ulema,* locally influential figures and the peasantry before guerrilla attacks could be launched. He seemed more inclined towards a mass movement than the vanguard-party framework of Hikmatyar. His moderate outlook and links with Sufi and literary networks in the north of the country earned Rabbani considerable support, particularly among the non-Pushtun, detribalized sections of society. Even in the Pushtun tribal areas, he made some inroads through local alliances, but his influence there was not comparable with the north.

The ideology of political Islamism did not prove a strong enough glue to keep the ethnically diverse groups together. With the widening gap in perceptions between the top leaders and competition to win support among the members of the Islamist constituency, the groups around Hikmatyar and Rabbani took different party names. While Rabbani retained Jamiat-i-Islami, the original title of the movement, Hikmatyar established Hizb-i-Islami.

Maulvi Yunis Khalis and Hikmatyar, both Pushtuns, worked together in the Hizb until 1979. Once again the question of strategy and approach towards the political and military issues resulted in a further split in the Hikmatyar group. Khalis preferred more and immediate guerrilla action after the communists captured power. Hikmatyar was never averse to the idea of a strong guerrilla movement but appeared to be more cautious—he probably wanted to build up an organized and trained force before plunging into the conflict. Khalis, much like Rabbani, is a highly accomplished religious scholar trained at Al-Azhar, who has worked in the traditional environment of the madrassah. Hikmatyar, on the other hand, is more of an Islamist intellectual and politician than an *alim* (religious scholar). To an outsider these differences may not appear sufficient reason for the fragmentation of the Islamist movement, but, however trivial, they have had a deep influence on group politics in Afghanistan. Khalis instead of forming a new group has continued to lead a break-away faction of the Hizb-i-Islami. These Islamist parties with identical names and programmes operate separately on both the political and the military front. Since 1985, the Jamiat and the Hizb of Hikmatyar have been locked in armed encounters inside Afghanistan.[17] The irony is that the Islamist parties have maintained large, disciplined and well-equipped forces, but have shown little restraint in fighting each other, largely to protect their own interests.

It would be pertinent to outline the politics, influence and capabilities of these parties in some detail.

JAMIAT-I-ISLAMI

The Jamiat has remained united under the leadership of Professor Burhanuddin Rabbani. Rabbani was inspired by the Islamist thought of Ikhwan-ul-Muslimeen when he was a student of theology at the Al-Azhar University during the 1966-68 period, where he actively participated in their programmes on campus. After returning to Kabul, he took up the position of Professor of Islamic Law at Kabul University and joined the Islamist group which had already been formed. He edited the *Mujalla-i-Shariat* (Journal of Islamic Law) until 1974, when he was ousted from the University for his involvement with the Islamic movement. The Daud regime ordered his arrest along with other Islamists but he succeeded in escaping to the border areas, where he attempted to organize the tribes against the regime. He remained underground in the country till 1978. After the communist *coup* he migrated to Pakistan. Rabbani is ethnically a Tajik from Badakhshan. He has developed broad-based support in the strategically important provinces of northern and northeastern Afghanistan. Ethnic minorities with strong anti-Russian sentiments like the Uzbek, Turkmens, and other primarily non-Pushtun groups have rallied around him. In Kunar, Ningrahar and Paktia, which are Pushtun provinces, Rabbani has struck some alliances with the *ulema* of traditional madrassahs.

During the war, the Jamiat guerrillas were active in Badakhshan, Takhar, Kunduz, Kapisa, Samangan, Parwan, Baghlan, and Mazar-i-Sharif.[18] The areas around the Salang pass and Panjsher Valley are strong bases of the Jamiat. This party has produced such legendary commanders as Ahmad Shah Masud, Zabiullah and Ismael Khan. Masud and Ismael set up a Council of Northern Areas where guerrilla activities were more co-ordinated than in any other part of Afghanistan. Masud used his stronghold in the Panjsher Valley to train guerrillas from other parts of the country—the guerrilla network developed by Masud has been one of the most effective and is comparable to any insurgency group in the world in strategy, tactics and leadership qualities.

Rabbani, head of the transitional government since 1992, has supported the democratic process for the enforcement of Islamic law.[19] He favours a free democratic order in society which would allow even the secularists to participate in the electoral process. Professor Rabbani has been open to establishing alliances among all the Mujahideen parties, and the Jamiat is an active coalition partner in the interim government. Professor Rabbani has pan-Islamic views, like the rest of the Islamist groups. He has tended to support the Iranian revolution and the unity of the Islamic *Ummah*. He has consistently urged the Islamic countries to support the Afghan resistance. He was not opposed to western economic and military assistance to the resistance via Pakistan, but argued that the Afghan resistance remain politically independent and capable of making its own decisions.

HIZB-I-ISLAMI (HIKMATYAR)

Gulbaddin Hikmatyar, the youngest of all the Mujahideen party leaders, is charismatic but is widely regarded by his critics as an extremist, hard-line and unaccommodating. Since its formation, the Hizb of Hikmatyar has preferred to maintain its independence from other groups, which in its view are loosely organized and were less effective as guerrilla organizations. Hikmatyar has not been charitable in condemning the role of traditionalist leaders, alleging they were part of the oligarchical system of the Afghan monarchy. He co-operated with them on a limited basis, but only under pressure from foreign friends and allies.

By social background, training and political experience, Hikmatyar is a puritanical Islamist. He has vowed to establish an Islamic Republic in Afghanistan more or less along the ideological foundations of Khomeini's Iran. Hikmatyar's radical Islamic credentials and his open praise of and support for the Islamic revolution in Iran helped him for a while to cultivate relations with the Islamic Republican Party, earned him enough good will to operate for some time a radio station in Iranian Balochistan and promoted some understanding with the Tehran-backed Shiite regional groups in Hazarajat. Relations with Iran have soured since Khomeini's death, though.

Hikmatyar's was a most effective guerrilla organization which operated on several fronts. He pursued a strict central-party line, exercised extreme caution in the recruitment of cadres, and laid stress on political training and modern organizational structure. The rigid party structure, political conformity and extensive intelligence network not only prevented infiltration by Kabul agents but also helped the Hizb to develop vital sources of information within the civil and military bureaucracy of the Marxist government. While a series of *coups* have been attempted against Najibullah by the military and civilian personnel of the regime, the abortive March 1990 *coup* led by former Defence Minister Shahnawaz Tanai had the co-operation of Hikmatyar guerrillas.[20]

Most of the prominent members of the Hizb are former students of Kabul University who worked with Hikmatyar in the Islamic Youth Organization. Educated youth from the detribalized middle class and defecting Pushtun soldiers and officers have preferred to join Hikmatyar. The availability of the disproportionately large economic and military resources allocated by Pakistan's Inter-Services Intelligence (ISI), along with an impressive organizational network and training facilities made the Hizb an attractive guerrilla party. More than the Jamiat, the Hizb has won wide support across tribal, ethnic and regional affiliations.

Hikmatyar's Hizb faction was probably the only group that effectively operated an urban insurgency in the major cities, particularly in Kabul. During the war, guerrilla operations by the Hizb were concentrated in Kunar, Laghman, Ningrahar, Kunduz, Kabul, Paghman, Parwan and Kapisa. Hikmatyar and his guerrillas operating inside Afghanistan were persistently accused of ambushing rival resistance groups for weapons and control of territory.[21] The Takhar incident, in which thirty commanders of the Jamiat were allegedly killed by Hizb guerrillas in the summer of 1989, considerably strained relations between the two parties.[22] The Hizb-Jamiat conflict was one reason why Hikmatyar decided to pull out of the interim government in February 1990.[23] It resurfaced in the form of a new power struggle, leading to armed clashes, following the collapse of the Marxist regime in April 1992.

Masud and Hikmatyar met in May 1992 and reached an agreement to end the fighting.[24] Reconciliation between these two parties will be essential for restoring stability, while their continuing confrontation will cause further fragmentation of the country.

HIZB-I-ISLAMI (YUNIS KHALIS)

Maulvi Yunis Khalis is an *alim* trained in the traditional madrassahs of Afghanistan, India and Pakistan. Despite his traditional Islamic background, Khalis was much attracted by the political Islamism of the Jamat-i-Islami and Ikhwan-ul-Muslimeen.

Before migrating to Pakistan in 1978, Khalis spent much of his time in teaching theology at his madrassah and writing extensively on religious subjects. Long before the launching of the Islamist movement, he published a religious magazine, *Payam-i-Haq* (Message of Truth), in 1961. He also founded a social reformist movement at about the same time. The focus of his writings was twofold: a) the transformation of the political system of Afghanistan in accordance with Islamic law and b) the supremacy of true Islamic values in society.

The Hizb-i-Islami of Yunis Khalis is a relatively small but well-equipped guerrilla organization. Khalis has personally commanded his fighters. In areas under his control, Khalis has attempted to establish an administrative set up.[25] He has a significant following in Ningrahar, Paktia, Kabul, Laghman, Kunar, and Wardak provinces[26] and has extended his influence to Qandahar. Jalaluddin Haqqani emerged as an effective commander of the Khalis group.

Most of the leading members of the Khalis group are students of Islamic madrassahs, *ulema,* tribal chiefs and intellectuals. Afghan army deserters have joined this party in large numbers. Khalis has favoured co-operation with the traditionalist groups. He has instructed his commanders and fighters to co-ordinate guerrilla actions with other parties.

The political views of Yunis Khalis on the Islamic reconstruction of Afghanistan are largely identical with those of Rabbani, Hikmatyar and Abdul Rasul Sayyaf.[27] The electoral process, an Islamic constitution and the evolving of a national consensus including the ethnic minorities are central to his political programme. Although he is greatly inspired by the ideological line of the Jamat-i-Islami, and admires its organizational work, he prefers to maintain a political distance. While Khalis insisted that the financial and material support for *Jihad* in Afghanistan should have come from the Islamic countries, which in his opinion was their religious obligation, he never rejected western aid through Pakistan.

ITIHAD-I-ISLAMI BARAYE AZAD-I-AFGHANISTAN (ISLAMIC UNITY FOR THE LIBERATION OF AFGHANISTAN)

This is rather a small group led by Professor Abdul-Rab Rasul Sayyaf. Itihad-i-Islami, or Islamic Alliance, was the name given to a loose coalition of five parties in January 1980. The coalition was created in much haste so that a representative of the Afghan resistance could attend the meeting of the Foreign Ministers of the Islamic Conference which was convened by Pakistan in May 1980 to protest the Soviet intervention in Afghanistan. Sayyaf was a compromise candidate to lead the coalition. Interestingly, when the Islamic alliance dissipated in 1985, Sayyaf continued to use the name Itihad-i-Islami for his group.

Sayyaf is an Islamist in his political outlook and was associated with the Islamist movement in Kabul, but he is better known as Wahhabi in Afghan circles. He has rather effectively used his Wahhabi credentials to secure large amounts of financial support from the Middle East, mainly through private sources. Lacking a political organization of his own, Sayyaf has generously distributed weapons and money among the semi-autonomous local groups. Perhaps he started out to build an organization by developing a clientele for money and weapons. Unfortunately Sayyaf never showed much care in selecting the recipients of his favours, which contributed to the emergence of independent groups without direction and control.

Another destabilizing element introduced by Sayyaf was the recruitment of Islamist fighters from conservative Arab countries.[28]

Arabs did not constitute a very large group, but their presence, estimated at one to two thousand, caused serious concern among other groups primarily because the Arab youth, highly motivated to kill or be killed and so drink the cup of Islamic martyrdom, was regarded as insensitive to the Afghan norms of warfare. On occasion they have slaughtered Afghan army defectors, thus discouraging further defections.

Sayyaf is a militant Islamist close to the radicalism of Hikmatyar. On many occasions he has described the objective of the Mujahideen resistance as not merely liberating the country but establishing an Islamic state.[29]

TRADITIONALIST PARTIES

The leaders of traditionalist Mujahideen parties, while subscribing to the principles of Islamic fundamentalism, consider the Islamist movement an alien ideology and markedly different from the Islamic ethos of the common Afghans.[30] The traditionalists contend that the religious views of the Islamists are not compatible with the popular faith of the masses. They argue that the Afghans have strictly followed the Islam of the mystic fraternities, and that the roots of Sufi networks run deep into society. There is considerable evidence for such claims. The descendant families of the established mystic saints popularly known as *pirs* (spiritual guides), and traditional *ulema* linked to the Sufi orders, have enjoyed tremendous social influence. Compared to the traditionalist Islamic influence, the Islamist movement was definitely weak prior to the launching of guerrilla resistance, but the war of resistance has entirely changed the conventional group identities. Under the pressure of the need for weapons, physical protection and social services, loyalties have become fluid. This is not to suggest that the traditional pattern of social linkages has completely broken down. Some of the old influences have weakened and new sets of elements have been introduced, but no society having undergone the devastation suffered by Afghanistan would be able to maintain a traditional balance of social forces.

The traditionalists have lost some of their influence by default. Despite a large following, the prominent *pirs* and traditional *ulema* did not recognize the need for political organization or political education of their adherents. Most of them took unqualified support for granted, and remained satisfied with the patronage system of the monarchy. They were more or less part of the Afghan establishment and members of the aristocratic social strata. More importantly, unlike the Islamists, the *pirs* and their tribal allies declared Jihad after the Soviet intervention.

The traditionalist party leaders have liberal political views. They have accepted secular democrats, nationalists and remnant elements of the old Afghan aristocracy in their parties. They do not appear to have a strict Islamic criteria for extending affiliations. Traditionalist leaders such as Gailani and Mujaddedi have in the past supported the notion of some role for Zahir Shah, if not the restoration of the monarchy, although some members of the traditionalist parties appeared unenthusiastic about the old monarch because he remained aloof from the Afghan Jihad.[31] Since the fall of the Marxist regime, traditionalist parties have formed a coalition with the Jamiat, and seem to have acquired a greater share in ministerial portfolios and high administrative positions. Mujaddedi was appointed as head of the transitional council, which he has continued to lead even after transferring power to Rabbani. Although some Islamic laws were enforced when Mujaddedi was the interim chief executive,[32] he favoured leaving the entire question of political restructuring to a *loya jirga*.

MILLI ISLAMI MAHAZ (NATIONAL ISLAMIC FRONT)

Sayyid Ahmad Gailani is the descendant of a spiritual family known as Naquib. He formed the National Islamic Front when he migrated to Pakistan in 1978. His traditional spiritual following and personal links with tribal chiefs, the land-owning aristocracy and Sufi families won Gailani's group massive support in Paktia, Ghazni, Paktika and Qandahar provinces. The Zadran, Mangal, Jaji, Ahmadzai, Tareen, Koochi and Sulemankhel tribes were his major support group, but even with impressive numerical support in large areas Gailani could not mobilize a large organized guerrilla force.

Gailani had strong personal relations with Zahir Shah and was one of the prominent members of the Afghan aristocracy. For the most part he remained informally associated with the monarchy and did not feel any compulsion to participate in electoral politics. It was only after the overthrow of Zahir Shah that Gailani, in co-operation with nationalists and tribal chiefs, organized a political party, Milli Watan Pak (National Homeland). The party stridently opposed the one-party rule which Daud imposed under the Hizb-i-Milli (National Party).[33]

Sayyid Gailani is pragmatic and flexible in his political behaviour. He even accepted an advisory position offered by Taraki, probably hoping that Khalq rule would be democratic and liberal. It did not take him much time to find out that the regime was too dependent on the Soviet Union and that its ideological rhetoric was divisive. He quit the post in just two months. Gailani's views on Islam and modern political issues are not dogmatic.[34] He represents a combination of religious leadership and traditional authority. Although he considers the pluralistic precepts of western democratic politics irrelevent to the political environment of Afghanistan, yet he believes that only a sort of representative government can unify the country. However, he has insisted all along that elections could not be conducted in the country due to the chaotic conditions, and instead has demanded that a traditional *jirga* be called to legitimize the Mujahideen government.[35] He is opposed to the dictatorship or exclusive dominance of one party. In his view, the existing guerrilla organizations could convert themselves into political parties in future elections.

Gailani strongly favoured the political and military support of the US and the free world against the former Soviet Union although at times he thought it was not enough, given their strategic interests in the region. He has received financial support mainly from the US and Saudi Arabia.[36] Since the ouster of the Marxist regime, he has supported the formation of a broad-based government in Kabul and has accepted the position of Foreign Minister.

JABHA NIJAT-I-MILLI (NATIONAL LIBERATION FRONT)

Sibghatullah Mujaddedi heads the transitional council which was formed in April 1992 to supervise the transfer of power from the falling Marxist regime. He belongs to the spiritual family of Mujadded Alf-Sani. Hazrat Shore Bazar of this family effectively mobilized the Afghan masses against the modernization reforms of King Amanullah in the 1920s. Mujaddedi is a recognized Islamic scholar and a follower of traditional Islam. He has been active in the Islamic movement since 1953. He spent four years in jail from 1955 to 1958 for his alleged involvement in a conspiracy to overthrow the monarchy. After the second Afghan elections of 1969 Mujaddedi organized the *ulema* of his sect under Ulema-i-Muhammad in 1970. The Ulema-i-Muhammad vehemently opposed growing Russian influence in Afghanistan and preached the solidarity of their nation with the Islamic world. After the 1973 coup, Mujaddedi actively opposed Daud's alliance with the *Parchamis,* fearing the ultimate domination of the communists over the state apparatus. He offered co-operation to Daud on condition that he disassociate himself from the communists.[37]

Mujaddedi has been affiliated with the Rabata-e-Alam-e-Islami (World Muslim League) which addresses the basic question of Islamic solidarity and co-operation among the Muslim states. When it became difficult for him to continue his political activity in Afghanistan, in 1974 he opted to lead the Islamic Centre of Scandinavia in Copenhagen. After the Saur coup he came to Pakistan and organized the National Liberation Front. The spiritual followers of his family, *ulema* of traditional Islamic sects and some tribal chiefs constitute his support base.

Mujaddedi is very much opposed to the Islamist political ideology, which in his opinion is foreign to the cultural base of Afghan Islam. Mujaddedi, like traditionalist Muslim scholars, views Islam as a creative relationship with those local values which more or less conform to the central principles of Islam. Thus, an Islamic system in Afghanistan would combine Islamic principles, Afghan traditions and such universal democratic values as are not repugnant to *sharia*. Mujaddedi's support for Islamic solidarity, social justice and self-determination for the people are central to his political programme. While leading a traditional Sufi network, he seems to be at ease with modern political thinking.

Although Mujaddedi is doctrinally opposed to the Islamists, he has partly extended co-operation to them. It was partly his moderate political views and conciliatory attitude that secured him a majority in the Afghan Majlis-i-shura (Consultative Assembly) which was convened to elect an interim government in February 1989.[38] He was elected head of the interim government, which Pakistan and other supporters of the resistance had hoped would attract wider participation. It did not, however, evolve into a pan-opposition representative institution, although it helped Mujaddedi and his traditionalist allies to gain better control of the foreign aid, which before had gone disproportionately to the Islamist parties. It was as one means of promoting unity among the resistance parties that Pakistan and the US decided to funnel arms and humanitarian assistance through the interim government. Mujaddedi's position as president, though recognized by only a few Muslim countries, definitely enabled him to gain stature and expand influence. His moderate views and the role he played as president of the interim government made him a convenient compromise candidate to head the transitional commission in Kabul in April 1992. In accordance with the Peshawar Accord, he transferred power to Rabbani in the last week of June 1992, but he continues to head the transitional council.[39]

HARKAT-I-INQILAB-I-ISLAM (ISLAMIC REVOLUTIONARY MOVEMENT)

Maulvi Muhammad Nabi Muhammadi, a traditional religious scholar from Logar, formed Harkat to lead a popular revolt against the *Khalqis* and their Soviet supporters. In the first few years of the resistance, Harkat emerged as a relatively popular and widely supported party. The personal standing of Muhammadi, particularly in the south, and his institutional links with the traditional *ulema* who run madrassahs, earned him broad support. His accommodating and somewhat liberal attitude attracted a large number of elements who could not otherwise fit into the rigid party structures of the

Islamists, but a lack of care in selecting allies horizontally across the ethnic and regional boundaries led to the decline of Harkat when better organized and more resourceful competitors entered the arena. While opportunists like Sher Agha, who for some time gained prominence within Harkat, defected to the Kabul regime, ethnic non-Pushtuns joined the Jamiat. Logar, which is Muhammadi's home province, remains his stronghold.

Muhammadi effectively used the madrassah network to build up political influence. It was a measure of his success that he was elected to the *Ulusi Jirga* (Parliament) in 1969. He won largely on an anti-communist plank. Outside and inside Parliament, Maulvi Muhammadi confronted the communists for their anti-Islamic views and links with the Soviet Union. He organized effective street agitation against the communists in 1970 to press the king to put a total ban on their activities and publications.

Liberal intellectuals, *ulema* and students of the religious institutions are the mainstay of Harkat's support. Besides his own kin groups from Andar, the Ghilzai, Mahmund, Hotak and Durrani tribes have supported Muhammadi. He has been open to alliances with the Islamist groups.

SHIITE PARTIES

Shiites, otherwise a small minority among the more numerous Sunni Afghans, constitute a composite majority in the central region, which is generally known as the Hazarajat. This area is the traditional homeland of the Hazara ethnic group. Qizilbash, who are mostly an urban and literate professional group, and some Orakzai Pushtun tribesmen also practice Shiite Islam. Largely owing to the geography of the Hazarajat and the concentration of the Shiite population in the region, Shiite groups have remained localized on their own turf. The fact is that they have not felt any strategic need to move out of their area or enter into alliances with the Sunni parties. The Hazarajat was practically abandoned by the Kabul-Soviet forces in 1981. It was an historic opportunity for the Hazaras to establish a sort of self-rule over their own territory, as they had resented being controlled by the largely Pushtun government from Kabul.

With the diminishing of Kabul's authority in their area, however, several Shiite groups have started to vie for power in the region. Three sets of actors have been actively involved in the making and breaking of groups. The balance of force among various groups has also been determined by these actors. They are: the Shiite religious establishment, the landowning class and the highly revered Shiite Sayyids, who claim family lineage from the Prophet of Islam. The land-owning class had traditionally dominated the political, economic and social scene of the Hazarajat until the founding of the local resistance parties. In the wake of the Iranian revolution, the Shiite religious establishment gained considerable political clout. The Sayyid families, in particular, took the initiative to organize resistance against the Kabul-Soviet forces. These developments, while not replacing the traditional influence of the land-owning class, at least brought to the fore new political forces.

Almost all these groups participated in the formation in 1979 of the first Shiite resistance group, the Shura-i Inqilab-i Ittifaqh-i Islami-i-Afghanistan (Revolutionary Council of the Islamic Union of Afghanistan). In composition, though broad-based, the Shura was dominated at the top by the Sayyid group. In a short time the Shura developed into a government with its own bureaucracy and military command which has been trained and led by defecting Afghan army officers. The Shura administration levied and collected taxes, promulgated its own laws and conscripted the Hazara youth to serve in the guerrilla force. The Shura's wide appeal and its ability to mobilize support was largely due to the issue of autonomy for the Hazarajat.

While the Shura ruled without any threat from Kabul, it faced local competition. As the corruption of the Shura functionaries and their excesses provoked resentment, new groups emerged to challenge its authority. A pro-Iranian Nasr (victory) party attempted to oust the Shura leadership of Sayyid Beheshti in 1982, ostensibly with the support of the radical Shiite religious scholars, some of whom had received education in Iran and were much influenced by the revolutionary thought of Khomeini. Educated youth and Hazaras who had lived in Iran supported Nasr. Daykundi has gradually emerged as the stronghold of Nasr and its headquarters.

In 1983 Sepah-i-Pasdaran (Army of the Guardians of the Revolution) was organized, mainly by clergy who had broken away from the Shura. Besides the claims of religious legitimacy, the leadership issue was central to the young clerics' disassociation from the Shura. Inspired by the decisively dominant role of the religious establishment in Iran, the young Hazara Sheikhs (religious scholars) sought greater recognition of their role in the politics of the Hazarajat. In opposing the power of the Sayyids over the Shura administration, Pasdaran received some co-operation from Nasr. Nasr and Pasdaran, though, controlled little territory—the radical clergy of these groups devoted resources more to effecting a religious social change than to leading military victories. Religious education, libraries and guidance are what the Sheikhs have pursued as the central goals of their movement.

Ayatollah Muhammad Asif Mohseni, a noted scholar and a native of Qandahar, organized Harkat-i-Islami (different from the Harkat of Muhammadi). He was first based in Qom in Iran, but later moved to Peshawar, Pakistan, after developing differences with the Iranian leaders. He has spent much of his life in education and teaching at the famous religious centre at Najaf in Iraq but he was very supportive in establishing religious institutions for the Shiite community of Afghanistan. Ayatollah Mohseni has not joined the Iran-supported coalition of eight parties (Wahdat). Although the Harkat of Mohseni has operated as a distinct group, mostly among the Sheikhs[40], it has tended to support the Shura administration. Ayatollah Mohseni is now an active member of the leadership council which formulated basic policy guidelines for the Islamization programme of the Mujahideen government.

The role of the Shiite resistance groups must be assessed on the basis of the ethnic and religious characteristics of the Hazarajat and of Iranian interest in the politics of this area. Hazaras have held the strong self-perception of being a subjugated minority. During the past ten years they seem to have acquired a political status which they never enjoyed in the past, when any attempt towards self-rule was crushed by the central government, and it might be difficult to deny autonomy to the Hazaras even if a unified and stable government emerges in the country. The Marxist regime offered recognition of the autonomy of the Hazaras in return for their participation in the national reconciliation. The Shiite groups declined these offers and continued identifying with the general cause of the Sunni resistance groups.[41] After the fall of the Marxist regime, the Wahdat-i-Islami, which is a coalition of eight Shiite groups, joined the interim government, having secured a fair share of power. The Wahdat had demanded a role in the decision-making process and autonomy for the ethnic minorities as conditions for its support to the Mujahideen government.[42]

The Shiite parties have received tremendous support from Iran during the entire period of the war of resistance. Tehran's interests lie more in promoting the unity of the Shiite parties[43] and seeking regional autonomy for the Hazaras. Alliance with the Sunni parties would hold only on the basis of greater representation in any future government and local autonomy for the Hazarajat. It is the first time that the Shiite parties have agreed to share power with the Sunni parties. A number of attempts by Sunni parties based in Pakistan to include Shiite groups in the interim government headed by Mujaddedi failed mainly because of the Shiites' insistence on being granted twenty-five per cent of the seats in the Majlis-i-Shura[44], a demand which most Afghans consider ridiculous due to the fact that the Shiite population is estimated at around fifteen per cent. Some even see an Iranian hand working to keep the Shiite resistance separate from the Sunni parties.[45]

POLITICS OF THE RESISTANCE

While the imperatives for transforming the Afghan resistance from a fragmented insurgency into a national liberation war were strong, the politically diverse Mujahideen groups preferred to organize their guerrilla forces separately. Intense factionalism within the resistance leadership was a constant source of disappointment to foreign supporters and internal commanders alike. Only under external persuasion combined with pressure and the incentives of economic aid and weapons did the resistance parties enter into loose coalitions and ineffective alliances. It was hard to discount the value of unity in bolstering the struggle, creating a better international image or even for political effect within the country, but under the impact of the scramble for power all the previous coalitions and alliances collapsed before gaining any respect or recognition.

Immediately after Khalq captured power in Kabul, some Afghan leaders in exile attempted to organize a broad-based united front, cutting across regional and tribal affiliations. In the summer of 1978 Burhanuddin Rabbani established the National Rescue Front, representing several religious groups and large numbers of former parliamentarians. Within less than two months the Front dissipated. In the autumn of 1978 Jamiat and Hikmatyar's Hizb-i-Islami agreed to form an alliance under the Harkat-i-Inqilab-i-Islami (Islamic Revolutionary Movement). It also lasted only a few months. By the end of 1979 hundreds of small and large resistance groups had emerged seeking recognition and support. Pakistan permitted only six groups to operate on its territory. This not only checked the further mushrooming of the groups and forced prospective leaders to join one of the recognized parties, but it also gave Pakistan leverage over Mujahideen politics by selecting which parties it would recognize.

The sentiment for unity appeared stronger than ever before when the Soviet Union invaded Afghanistan. The Islamic countries who were offering aid counselled the Afghan parties to establish a coalition. Five parties led by Rabbani, Khalis, Gailani, Muhammadi and Mujaddedi agreed in January 1980 to form the Islamic Alliance for the Liberation of Afghanistan.[46] Hikmatyar refused to join the alliance for two reasons. First, he objected to accepting Gailani and Mujaddedi, whom he accused of being opportunists, in the fold of the alliance. Second, he demanded a larger role in the alliance, claiming that his Party was the most effective inside Afganistan. The alliance held together for less than six months. The forthcoming Islamic Foreign Ministers' Conference in May 1980 was the reason that the Afghan groups put up a semblance of unity. It was expedient and necessary to promote this image for winning the support of the Islamic countries and attracting foreign aid. The alliance, however, did not succeed beyond making an appearance before the conference. The Islamic foreign ministers' meeting did not accord diplomatic recognition to the alliance. The Iranian government adroitly made the head of each of the parties in the alliance, including the leaders of the Tehran-based parties, part of its delegation to the conference. Within months after the conference, the alliance withered away. Rivalry between the Islamist and traditionalist groupings caused the rupture.

In June 1981 the traditionalist parties formed a new alliance, called the United Islamic Front of Afghanistan.[47] The Front brought together Milli Islami Mahaz led by Gailani, Jabha Nijat-e-Milli led by Mujaddedi and Harkat led by Muhammadi. The traditionalist front was created with the objectives of co-ordinating the political and military activities of the member organizations, which they thought was necessary to win recognition and enhance their fighting capabilities. They drew up a comprehensive charter outlining a common organizational structure, political aims and strategy. The Front set up a joint administrative machinery which functioned under sixteen committees and bureaux. The leader of the alliance, who was elected for a one-year term, was designated president, while the leaders of the member parties acted as vice-presidents. The leaders of the alliance determined joint positions on political issues and military strategy, but the military committees of the component organizations never merged into a single command structure. The traditionalists in principle supported the formation of a broader political alliance with the Islamists on the pattern of a united front strategy.

The Islamists formed a rival alliance, the Islamic Alliance of Afghan Mujahideen, in August 1981, a few months after the traditionalists organized their united front. The Islamist Alliance, which was extended in March 1982, represented three major and four small organizations. The seven members of the group were as follows: the two factions of Hizb-i-Islami, one led by Hikmatyar and the other by Khalis, Jamiat-i-Islami led by Rabbani, two break-away factions of Harkat-i-Inqilab-i Islami led by Maulvi Nasrullah Mansur and Maulvi Rafiullah Muzeen, Ittihad-i-Islami Baraye Azadi Afghanistan led by Sayyaf, and Jabha Nijat-i-Milli led by Maulvi Muhammad Mir, who had defected from Mujaddedi's party. A sixty-member advisory council with equal representation was created by the Alliance as a common forum for discussion and the formulation of joint policies.[48]

The leadership of the alliance rotated on a monthly basis among the member parties. The Islamist Alliance appeared to be developing into an integrated party structure when in May 1983 it took two important decisions: the member parties reaffirmed their willingness, first, to merge into a single organization and second, to fight under one command. None of these decisions could be translated into practical steps though, and with interpersonal and intra-group rivalries and intermittent fighting between the Hizb and Jamiat, unity remained as fragile as ever.[49]

In 1985, under pressure from the internal commanders who confronted the escalation of the Soviet war, and the greater degree of influence exercised by Pakistan, Saudi Arabia and the western supporters, all the seven major parties, earlier divided into rival alliances, formed a larger coalition, the Islami Ittehad-i-Mujahideen-i-Afghanistan (Islamic Unity of the Afghan Mujahideen). The seven leaders of the parties comprised a Supreme Council which took all the important decisions on the basis of unanimity. A number of committees with equal representation of the member parties were set up to co-ordinate policies. The leadership of the coalition rotated among the parties. The leader was designated spokesman rather than president. Collective leadership was the idea behind the formation of the supreme council, and the principle of unanimity and rotation of the spokesman for a limited period indicated the same. Apparently, the common objectives of evicting the Soviets from the country and defeating the regime set up by Moscow, along with financial considerations, kept the internal squabbles suppressed. Pakistan's General Ziaul Haq personally intervened whenever it was necessary to keep the unity of the Mujahideen intact.[50]

As the Geneva talks on Afghanistan progressed, Pakistan's interest in maintaining the Mujahideen coalition grew stronger. Pakistan had been adamant on the formation of an interim government to take place simultaneously with the departure of the Soviet forces from Afghanistan. Islamabad envisaged a transitional government shared by the Mujahideen coalition and those internal elements not closely associated with the Kabul regime.[51] In the final round of the Geneva talks in April 1988, however, Pakistan, under much US pressure, compromised its long-standing position on the interim government.[52]

Perhaps Pakistan and the United States were too optimistic about the military victory of the Mujahideen at that time, and thought that the Marxist regime would collapse anyway.

When the Soviet military withdrawal from Afghanistan reached its final stages, the seven Afghan parties in the IUAM, with considerable assistance from Pakistan and other foreign supporters, convened a Shura to elect an interim government that would take control in Kabul if the Soviet-backed regime fell. The idea was to form a broad-based coalition of the Afghan opposition as a political alternative to the Kabul regime. It was not a new concept, as we have noted earlier. A month before the conclusion of the Geneva talks, the Supreme Council of the IUAM passed a resolution setting up an interim government with Engineer Ahmad Shah as its president and Zabihullah Mujaddedi as vice-president.[53] The decision to establish an independent council or Shura simultaneously with the withdrawal of Soviet forces was contained in the same resolution. Even before this, a proposal to form a government-in-exile had been discussed on several occasions, but it was killed each time due to the lack of agreement on representation for each group.[54]

Pursuant to an earlier decision of the IUAM, the Mujahideen of the Afghan Shura met in February 1989 in Islamabad to elect the interim government. About 425 delegates participated in the Shura proceedings. Given the numbers, it was an impressive gathering, unprecedented in the history of the Afghan resistance, but as soon as deliberations on the agenda started, the Shura became deadlocked both on procedural as well as on substantive issues. The session lasted more than two weeks. Neither the Afghan resistance nor the Pakistani hosts wanted to end the Shura without electing the interim government. After much bargaining and outside pressures the Shura settled the procedural questions, which were significant as no constitution existed other than a rough charter drawn up by the IUAM.[55]

After rejecting the nominations of Engineer Ahmad Shah and Muhammad Nabi Muhammadi for the offices of president and prime minister respectively, the house adopted a novel method of election. It established a ranking system for the elective officers, with president at the top followed by prime minister and other cabinet posts. The members of the Shura voted for the party leadership in a single ballot. According to the principle established, the highest vote-getter would become the president, thus apportioning the offices in accordance with the rank of each party as shown by the number of votes cast in its favour. Professor Sibghatullah Mujaddedi securing the highest votes was elected as president and Abdul Rasul Sayyaf as the prime minister.[56] This procedure enabled all the member parties to have a ministerial position in the interim government.[57]

Although an interim government was formed, it lacked the necessary broad-based support. A great majority of the internal commanders either were not invited or abstained, resenting the pressures the Pakistan government was exercising in shaping the interim government. Although the Shiite parties based in Iran came to participate, they returned without doing so as their demand for higher representation on the Shura was rejected by the Sunni parties. None of the exiled figures in the west nor any prominent individual from within Afghanistan other than the resistance leaders was invited to join the Shura. For this very reason, Pakistan, which had been so enthusiastic in support of the interim government, politely refused to extend recognition, saying it was 'premature.'[46] Washington also reacted cautiously, waiting to see whether the interim government could establish effective authority in Afghanistan.[59] Pakistan and the United States wanted the interim government to capture Khost or Jalalabad before they would accord recognition—this can be deduced from the Mujahideen assault on Jalalabad following the formation of the interim government.

After its establishment, the interim government remained handicapped. Firstly, it failed to convince the Shiite parties to join it. Several rounds of talks did not break the impasse on the representation issue. Secondly, internal dissension prevented the government from growing in stature. Finally, it lost whatever legitimacy it had in February 1990 as it was elected only for a year. Efforts to reconstitute the Shura did not succeed. In practice it ceased to exist when Hikmatyar's Hizb-i-Islami withdrew its support of the interim government, partly because of its disagreements with the Jamiat-i-Islami and partly because its one-year tenure had lapsed.[60]

GUERILLA STRATEGIES

Guerrilla warfare has proved an effective strategy for causing disproportionate damage to superior forces. The dispossessed people, both in wars of national liberation as well as in revolutionary wars, have amply demonstrated that guerrilla strategy is the appropriate choice of the weak in challenging the strong. In this sense, the ends and means of guerrilla warfare largely differ from the purposes and operational doctrines of conventional conflict. More than territorial conquest, which would naturally involve direct and decisive confrontation with the enemy forces, guerrilla strategy emphasizes harassment and surprise to destroy enemy morale and to force him to recognize that the costs of continued war are too high.

In Afghanistan, armed struggle has historically been the conventional choice of society for bringing about desired political change in Kabul, or seeking the end of foreign domination. The use of force against a hostile group or entity is a widely practiced and accepted norm among the Afghans in a cultural and religious sense. In this respect, the popular uprising against the Kabul regime and the sustained insurgency against the Soviet occupation has not only been symptomatic of traditional Afghan behaviour but is also a true reflection of the nation's determination to free the country of foreign aggressors and internal usurpers.

Although Afghan society was fragmented, it survived the extermination campaigns persistently launched by the Kabul-Soviet forces. Two factors in particular aided the evolution of the Afghan resistance and its rapid growth. First were the mobilizing effects of Islam and nationalism. The Marxist regime was handicapped by the political burden of an alien ideology and was isolated and vulnerable to Mujahideen propaganda that it had soiled Afghan honour by subjecting the country and the population to Soviet military intervention. Images of the regime functionaries as quislings and communists provoked resistance. On the other hand, the goal set by the opposition group—to rid the country of foreign domination brought about by the Afghan usurpers—had both cultural and religious credibility.

Secondly, a national uprising in the traditional pattern of tribal revolt created a natural psychological and political milieu for armed resistance. The mass mobilization, which was instantaneous, in fact preceded the attempts at political organization and the emergence of leadership. This was quite in contrast to the typical situation in the Third World, where the dormant masses have to be educated around a national cause and motivated to participate in the guerrilla war. In the case of Afghanistan, leadership potential existed in the mass revolt that inspired prominent social, religious and political figures to convert the unrest into popular resistance. The degree of mass support compensated somewhat for the absence of critical elements in the development of a national resistance, e. g. an effective political organization or a charismatic figure to rally around.

The level of mobility and survivability achieved by the Afghan resistance was possible because of the political isolation of the Marxist authorities and the alienation of the population. The Afghan guerrillas had access to a population that was willing to take risks in assisting the insurgency by making available food, shelter and necessary information.[61] The genuine and mutually supportive relationship of the resistance with the local population offset the awesome power of the Kabul-Soviet forces. Undoubtedly, it was their failure to win the masses to their side that compelled the successive Marxist regimes, guided and aided by the Soviets, to target the population and its means of sustenance. This further promoted hostility towards the regime.

The guerrilla strategy of the Afghan resistance was aimed at achieving a combination of political and military objectives. The following were the goals central to their armed struggle:

a. denial of legitimacy to the Soviet-backed Marxist regimes, and perpetuation of their moral and political alienation from the population.
b. establishment of a guerrilla infrastructure and parallel administrative control in the liberated zones.
c. maintenance of military stalemate through a war of attrition.

The military strategies of the Afghan guerrilla organizations were generally successful in denying the Kabul regime stable and effective control of the countryside, but because of the divisions within its ranks the resistance could not produce the desired political effect which it might otherwise have done by forging unity. Although a preference for maintaining separate organizational networks did not affect their efforts to mobilize and maintain support in the respective tribal, ethnic, and ideological constituencies, it nonetheless presented serious problems of strategic co-ordination on various fronts inside Afghanistan.

The military tactics of the Afghan resistance evolved from tribal revolt patterns to modern guerrilla warfare. Vast combat experience, training skills, external support and the availability of modern weapons contributed immensely to the development of Afghan guerrilla warfare. Initially, the isolated revolts took the shape of tribal war. The Afghans launched unorganized frontal assaults against government outposts in large numbers. Direct confrontation with a militarily superior enemy inflicted heavy casualties on an untrained and poorly equipped rural population. Though he had a gun and knew how to use it, the Afghan tribal warrior was unaware of the awesome killing power of modern arms and the organizational strength of the Kabul-Soviet troops. The spirit of individual heroism and the flamboyant tribal style were ineffective in achieving military success.

Over the years, however, the Afghan resistance did achieve remarkable success in two critical areas of guerrilla warfare. First, it gradually created skilful cadres to organize and lead local fighting units. In all respects, it fielded modern and capable local leaders who were trained and experienced enough to multiply their force efficiency by carefully selecting the targets and mobilizing appropriate numbers. Second, some of the Afghan resistance parties and internal commanders developed a regional system of alliances across ethnic boundaries and tactical co-ordination in strategic areas. This pattern of resistance was of great significance for increasing cohesion and conglomeration among the diverse ethnic and regional groups. The resistance in the northern provinces of Afghanistan achieved better strategic co-ordination than anywhere else in the country. There, commander Ahmad Shah Masud of the Jamiat-i-Islami was the guiding spirit behind co-operative guerrilla networks.[62] Using his Panjsher stronghold, Masud provided training facilities to fighters from other areas and showed the way to other commanders for operating a local administration. In almost in all the regions of Afghanistan capable guerrilla commanders emerged.[63] They learned to coexist, and quite often set up joint commands for specific operations. This process promoted a broader coalition of internal forces in the north, if not a common command and control structure.

The guerrilla operations of the Mujahideen strongly emphasized the role of sabotage in producing political and psychological effects. The harassment of government forces was an effective means of reducing their credibility and establishing control in the countryside. The value of sabotage lay in undermining the authority of the superior forces. Also, targeting the physical destruction of facilities considered directly or indirectly helpful to those in power seriously limited their mobilization capability.[64] The sabotage of the strategic assets of the Marxist regime, such as ammunition depots, aircraft and other military gear, was largely an 'inside job' as the Afghan military services were infiltrated by the Mujahideen. The destruction of about twenty Afghan Mig-21 jet fighters at Shindand air base in north-west Afghanistan in June 1985 was one of the more spectacular acts of sabotage. It virtually wiped out a fourth of the Afghan military's fighter planes.[65]

The well-planned and organized sabotage operations were aimed mostly at the public utilities. The electrical generating and distribution lines were the major targets. The frequent disruption of electric supply to the major cities demonstrated to the urban population that the Marxist regime had no control outside its fortified bases. The attacks on communication networks equally impaired the effectiveness of the regime. The destruction of bridges, mining of roads, and destruction of telephone lines were favourite tactics of the Mujahideen. The acts of sabotage proved increasingly demoralizing for the government soldiers. They could not escape the fact that they were fighting a popular insurgency. The element of surprise in the sabotage tactics, and its effectiveness in hindering the counter-insurgency operations of the Kabul forces, made it a popular means of waging guerrilla war in Afghanistan.

The second major military tactic of the Mujahideen was ambush. The supply columns of the Kabul-Soviet forces were a common target. Attacks against supply lines served a dual purpose. They were a source of weapons, and they also limited the combat capabilities of forces located in remote areas. The siege of outposts and night-time rocket attacks became common. The Mujahideen changed their early strategy and avoided running down the enemy posts, as this had proved costly in terms of the loss of lives. Such attacks were launched for the purpose of harassment. However, if the Mujahideen intelligence indicated success, they tried to capture the post and take away the arsenal. Airfields were also the target of rocket attacks, particularly the Bagram airbase, which provided logistic support to the helicopter gunships and also served as an operational base. Although attacks on the Bagram base or on other bases on the eastern and southern fronts never reduced the effectiveness of the helicopter gunships, they did demonstrate the ability of the guerrillas to penetrate into strategically sensitive areas.

Though the Afghan resistance was based in rural areas, some of the groups maintained successful underground networks in the cities. The urban form of the resistance centred on two types of activities: mass demonstrations and terror attacks. Using *shabnamas* (night letters), pamphlets and posters, the urban guerrilla networks called for strikes and street agitation. Such campaigns of protest were more paralyzing and common in the early years of the war. Street demonstrations became fewer with the tightening of Kabul's security and the use of excessive violence by the security forces.

Terror attacks were practised by almost all guerrilla groups but Hikmatyar's Hizb-i-Islami was the most effective. The commonest form of urban terror was assassination. These operations of the Afghan Mujahideen, more frequent in 1981, were careful and selective.[66] The guerrillas targeted workers of Khalq and Parcham, high ranking officials of the Marxist regime and Soviet army and civilian personnel. The resistance avoided indiscriminate killing of the pro-regime elements. The tribal character of Afghan society makes wide-spread terror against individuals counter-productive. The physical elimination of a village chief carried always the potential to turn the whole village against the guerrillas. The assassination of officials of the regime and party workers was meant purely to maximize the psychological effects of guerrilla warfare. It prevented the regime from building popular support among the masses.

Taking advantage of the retreat of the Kabul-Soviet forces from areas which they found hard to defend, some of the Mujahideen groups built up fortified bases. The move was partly inspired by the survival of commander Ahmad Shah Masud's base networks in the Panjsher area. In 1985 the guerillas dug an underground base in a narrow valley near Zhawar in Paktia province, not far from the Pakistan border. It was a mile-long complex of man-made caverns ringed by fortifications, and defended by tanks, artillery and anti-aircraft batteries. This base attracted world-wide attention for its facilities ranging from shelters for captured tanks to hospital and training facilities.[67] For quite some time the garrison was considered safe and easy to defend. As this guerrilla centre gained prestige as a main communication and supply base, the Kabul-Soviet forces launched a major offensive in 1986 and finally destroyed it, inflicting heavy casualties on the guerrillas.[68] The local commanders blundered in underestimating the power of the Soviet forces to destroy them. Their confidence had grown as they survived sporadic aerial and artillery bombardments. Furthermore this type of territorial objective, in the face of the damaging fighting capability of the enemy, negated the fundamental doctrines of guerrilla warfare.

The acquisition of better weaponry contributed to the excessive self-assurance of the Mujahideen. A radical change occurred in their approach and effectiveness with the significant inflow of heavy equipment for anti-tank and anti-aircraft warfare. As it became clear that the Soviets were bogged down in Afghanistan, countries sympathetic to the guerrillas increasingly made available modern weaponry. Better arms, such as the coveted AK-47 Kalashnikov assault rifle which was used by the Soviet forces, mortar launchers, heavy machine guns and mines vastly improved the lethality of guerrilla attacks, but even with the proliferation of new weapons, the basic problems confronting by the resistance remained. The supply of weapons was neither enough to sustain attacks nor evenly distributed among the various fronts. Corruption in the supply line and the sale of weapons by Mujahideen groups were frequently reported. On the whole though, the variety of weapons captured from the Kabul forces and those shipped by foreign supporters made the Mujahideen more formidable on the battlefield than their traditional British-style Enfields or weapons manufactured in the markets of Darra in Pakistan could have done.[69]

Although the Afghan resistance maintained surprise and initiative, it remained vulnerable to Soviet helicopter assault operations until 1986, when Britain supplied Blowpipe anti-aircraft missiles, followed soon after by Stingers from the US. Two factors were central to furnishing the resistance with air-defence weapons. First was the effectiveness of the Soviet helicopter-gunships, which gave them freedom to choose the time and place of their attacks against the Afghan resistance. In 1986, the Soviet counter-insurgency campaigns were more aggressive, inflicting heavy losses on the Afghan freedom fighters. Second was the western desire to escalate the war in Afghanistan, forcing the Soviets to pay a higher cost for their intervention. Before the transfer of shoulder-fired Stinger missiles, the Mujahideen relied on the Swiss Oerlikon 20mm anti-aircraft cannon, which was the largest air-defence system in their hands, and from 1982 they had the Chinese Type-75 14.5mm anti-aircraft machine guns, besides SAM-7 clones. Before the availability of Blowpipes and Stingers, no more than twenty aircraft per year were shot down by the Mujahideen.[70] What made the Stinger a dangerous weapon was its kill probabilities, which US test results put at seventy-seven per cent. Operated by less skilled Mujahideen crews the kill ratio of the Stinger never touched this percentage,[71] but even according to more conservative estimates of a kill probability of thirty-three per cent, the Mujahideen were able to destroy 270 aircraft per year after 1985.[72] In addition, operational attrition caused many Soviet aircraft losses.

The daunting technology of the Stinger made it a big hit with the Afghan insurgency.[73] It not only restored Mujahideen morale, which had been badly hurt by the Mi-24 'Hind' gunships, but also changed the course of the war. The hitting power of the missiles severely restricted the mobility of the Soviet counter-insurgency squads. Given the difficult terrain in Afghanistan and the large numbers of troops required, the Soviets were unwilling to take risks in deploying men on the ground to carry out punitive or search-and-destroy missions. The Soviet position became more painful with the escalation of weapons technologies transferred to the Afghan insurgency by t West. Consequently, the Soviets, tired by the stalemated war, a in line with domestic concerns and new international initiativ decided to seek a negotiated withdrawal rather than to expand t war.

THE BATTLE OF JALALABAD AND AFTER

On 8 March 1989, three weeks after the Soviet troops complet their withdrawal, some groups of the Mujahideen launched a fron assault on the city of Jalalabad. In the first two weeks of the campai the Mujahideen overran the heavily-defended Afghan Army garris at Samarkhail and completely demolished the outer defence lin but the attack soon bogged down and developed into a long sieg The battle of Jalalabad was the first determined effort by the Afgh resistance to capture a major city. The failure to take this key easte city, which had been widely predicted to fall in the absence of t Soviet forces, raised fresh questions about the military option break the stalemate.

The decision to capture Jalalabad was motivated by two importa considerations. The first was to provide a territorial base to the new formed interim government, as without the government establishi effective authority inside Afghanistan, none of the foreign supporte was willing to accord it recognition. Second was the calculation th the fall of a major urban centre to the resistance would cause devastating blow to the Soviet-backed government. Much of t optimism in the guerrilla ranks was fed by the expectation that t government defences in the city would crumble from within. To t surprise of the resistance and its backers, the Afghan governme forces repulsed the guerrilla attacks and finally broke the siege b clearing the Kabul-Jalalabad highway. Even though besiege from all sides for about four months, the defenders held o with air-dropped supplies. The move to capture Jalalabad turned o to be a fiasco for the Afghan guerrillas. The outcome of t battle immensely affected the psychological and politic environment of the Afghan conflict, raising important questions abo e organization and guerrilla strategies of the resistance. Why was e decision to attack Jalalabad taken so hurriedly? Were all the sistance groups taken into confidence? What kind of tactical sjudgments were made? Had the Afghan guerrillas attained the vel of organization and co-ordination necessary to capture a city? hat changes did the Jalalabad setback bring to the strategic thinking the Afghan resistance? These are some of the questions which ed to be examined in the context of the military stalemate that evailed in Afghanistan until the breakup of the Soviet Union, the ects of which shook the Marxist regime in Kabul to its foundations.

With hindsight it appears that the decision to take over Jalalabad as ill-advised and premature.[74] Although a few Mujahideen groups th local influence had maintained their presence on the outer rimeters of the city long before the full-scale attack began, they ere not wholly prepared to confront the Kabul government fences. The planning to execute the offensive ignored vital facts out the Afghan resistance. The guerrillas, trained and experienced hit-and-run tactics, were ill-equipped to fight a set-piece battle. odern conventional wars, with the greater kill-power of high-chnology weapons, are vastly different from previous conflicts volving guerrilla armies and state forces. The Afghan resistance d yet to learn how to fight a conventional war to gain control of e cities. Therefore, by abandoning the countryside, the Marxist gime did survive and continued to claim legitimacy even though control was limited to major towns and military garrisons.

Another important factor responsible for the Jalalabad debacle as the lack of co-ordination and of a unified command structure. s pointed out earlier, the resistance groups kept their separate ommands, and showed no inclination towards a merger on a tional or even regional level by combining military operations. It perplexing to a military mind to suggest that the transformation of e fragmented guerrilla organisations into a conventional army ould be possible, unless they agreed to set up a common military ontrol system. The failure to block all Kabul's supply routes to lalabad at the time of the final assault might be attributed to a lack co-ordination among the guerrilla networks. Supplies and inforcements from Kabul continued to reach the city until 20 March when the guerrilla advance toward the airport had considerably slowed.[75] Moreover, not all the Mujahideen groups were mobilized to send their guerrillas to reinforce the attack. A few commanders joined in a bid to save the situation as enormous prestige was attached to the effort[76], but their intervention was too late to produce any military results.

The Mujahideen assault force was unable to deal with fixed defences, minefields, and dug-in tanks, mainly because of poor training and light equipment. Generally exposed and concentrated in static positions, the guerrillas were vulnerable to lethal bombing by MIG-21s and SU-17s which stayed out of range of the Stingers by flying at high altitude. Scud missile and rocket attacks complemented the high explosives delivered by the bombers. The Mujahideen had very little in terms of heavy equipment, organization, and quick reinforcements to overrun the strong Kabul government defences.

Even the size of the Mujahideen assault force, which at its peak never exceeded eight thousand, was numerically too small to defeat a well-entrenched, heavily equipped and disciplined regular army of about 12-15,000. The logistical requirement in both men and material of those laying siege and planning to take over a city requires a ratio of at least 3-1, if not more.[77] Despite the fact that the Mujahideen had gained control of the 80-kilometre Torkham-Jalalabad highway, they could not supply themselves with enough weapons or mobilize a force-level sufficient to sustain the initial momentum.

The stalemated battle around Jalalabad drastically changed the perceptions of both the Marxist regime and the Afghan resistance. The Mujahideen in their state of disunity and uncoordinated struggle were unprepared to destroy the military and political machine of Najibullah. The Marxist regime, even with enormous Soviet military hardware, was unable to crush the resistance. Najibullah was able to make some political gain out of the successful defence of Jalalabad by showing that his regime could survive without the direct involvement of the Soviet Union. This assessment, widely peddled by most of the regional and western journalists, considerably lowered the stock of the resistance. It would be wrong to assume, however, that the military threat from the Mujahideen totally disappeared or that they were crippled enough not plan other major conventional offensives. The Mujahideen held on to their positions around Jalalabad and continued to control wide areas between the city and the Pakistan border in the east.

The Afghan resistance groups, divided as they are even after replacing the Marxist regime, have had different strategies for toppling Najibullah. The difference between the approaches of the Hizb-i-Islami of Hikmatyar and the rest of the parties in the interim government became pronounced after the *coup* attempt in March 1990 led by General Tanai.[78] While Hikmatyar supported the *coup,* all six of the other party leaders refused to join his alliance with Tanai. They characterized the event as a fight between the two communist factions.[79] Following the failed *coup,* a new alliance between diverse elements among the former high ranking Khalq officials, defecting army generals, Hikmatyar and some internal commanders took shape. This coalition, which blurred the old ideological cleavage, was based on the conviction that a *coup* from within the Kabul regime was a more realistic option than attacking the cities.[80] In line with this strategy, they preferred to concentrate around Kabul and increased links with the likely *coup* plotters in order to exploit events quickly. Battles over the cities, in their assessment, would be costly in human terms and devastating. With the presence of scores of experienced Afghan army generals and other high ranking military officers, mostly of Pushtun origin[81], this coalition attempted to mould the guerrillas and the defecting soldiers into an army under one command.

On the other hand, the Afghan Interim Government (AIG), which was receiving the bulk of the supplies and appeared to be favoured by Washington, had different plans. It preferred the tightening of the guerrilla grip on the besieged provincial cities and taking them over before launching a final assault on Kabul. In eastern Afghanistan, guerrilla groups led by various parties of the AIG, notably the forces of commander Jalaluddin Haqqani, launched a new offensive around Khost in May 1991. Khost was already besieged and was being supplied only by airdrops. After the Jalalabad offensive, this town was the first to fall to the Mujahideen.[82] In a reappraisal of their strategy, the resistance parties planned multi-pronged co-ordinated attacks against more than one target. Their aim was to spread the Kabul forces thin by opening up diversionary fronts and to neutralize the regime's effective use of air-power. Such a strategy, though, required a large number of well-organized and better-equipped guerrilla armies, which the resistance found beyond its means to field.[83]

SANCTUARIES AND INTERNATIONAL SUPPORT

The sanctuaries in neighbouring Iran and Pakistan and military assistance from a number of countries, particularly Saudi Arabia and the United States, were central to the effectiveness of the Afghan resistance. No matter how socially broad-based, the Mujahideen might not otherwise have survived the protracted conflict against the vast kill-power of the Kabul-Soviet forces. Pakistan assumed a crucial role in extending protection to the guerrilla leaders, letting the resistance parties organize on its territory and allowing foreign-supplied weapons to cross into Afghanistan.

Pakistan's open defiance of the Soviet move in Afghanistan was encouraged by US support. The Soviet role in regional conflicts in general and the direct military intervention in Afghanistan in particular worsened superpower relations. Washington viewed the Soviet invasion in a broad strategic context, as an example of Soviet willingness to use its growing military power beyond its borders. The Soviet posture was interpreted as threatening not only American interests but also peace, stability and freedom in the areas adjacent to the Soviet Union. The US response was guided by two concerns: first, that American inaction would generate the impression at home and abroad that it lacked the political will to counterbalance the Soviet threat in the Third World, sending wrong signals to the Soviet Union, and second, there were apprehensions that the pro-West allies in south-west Asia who saw the Russian foray into Afghanistan as directed at them politically would cut a deal with Moscow. The image of the United States as irresolute and undependable in the face of Soviet aggressiveness might have further eroded American influence in the region, where it had already suffered a heavy blow from the Iranian revolution.

The US response was also shaped by fundamental changes in its foreign policy. First was the new image of the Soviet Union as an expansionist and destabilizing power seeking domination. This change of perception was central to the new foreign policy of the Carter administration.[84] Second was the growing consensus within the United States for a stronger defence posture.[85] Both to assuage the sense of vulnerability among its allies in the region and to signal to the Soviet Union the American resolve to protect western interests, the Carter administration took a number of measures. Among these was the statement, known as the Carter Doctrine, that the United States would defend its interests in the Gulf region, if necessary by military means. To make this stance more firm and credible, the creation of a rapid deployment force was announced, but no military counter-pressure was initiated. However, Washington called for punitive action in broad areas, thus indicating that the superpower detente had collapsed. This included a limited grain embargo, a ban on sales of high technology to the Soviet Union, in which Japan and the European allies were strongly urged to join, a restriction on low interest, government-guaranteed credit for the Soviet Union, the Olympic boycott, and the diplomatic condemnation and isolation of Moscow. President Carter also advised the Senate to defer debate on the ratification of the SALT II treaty, having become convinced that the Senate would never approve the treaty while Soviet troops remained in Afghanistan.

More specific to Afghanistan, US policy centred on the covert arming of the Mujahideen. The Mujahideen had established organizations and strategy well before the US become involved, and required only military assistance.[86] Washington's approach was cautious as it did not want to escalate the level of conflict, fearing that that would precipitate the introduction of more Soviet troops and endanger Pakistan's security. At first, the US and Saudi Arabia provided about $30 million each for the purchase of Soviet-style weapons manufactured in Egypt and China.[87] As the low-intensity conflict began to bog down the Soviet forces, US support to the guerrillas increased significantly. The objective was to exact a heavy price from the Soviet Union for its invasion and prevent Afghanistan from becoming a platform for further expansion.[88] Although the

Reagan administration chalked out a more comprehensive strategy for assisting anti-communist guerrilla movements, rolling the Soviet Union back from Afghanistan solely through resistance was never considered a realistic option. Rather, the United States pursued a multi-track policy, of which equipping the resistance was one element. This was premised on the assumption that raising the Soviets' costs in Afghanistan and frustrating their attempts to pacify the country would make the Kremlin seek a negotiated withdrawal. Therefore, direct negotiations with the Soviet Union in the context of easing superpower tensions and resolving Third World conflict figured equally importantly.

From the US and Pakistani perspective, the overall settlement of the Afghanistan conflict hinged on the Soviet withdrawal and the right of the Afghan people to exercise self-determination. Achieving these two objectives required US-Pakistan security co-operation on the one hand, and the raising of the level of resistance on the other. The latter decision was taken after pressure from the US Congress, which appeared more sympathetic to the Afghan cause than to any other anti-communist resistance. Over the years, Congress approved more than $2 billion for the Afghan resistance, about $400 million annually on an average in the years preceding the Soviet withdrawal.[89] The supply of sophisticated anti-aircraft weapons following 1986 was instrumental in blunting the Soviet bombardment of Mujahideen positions. Washington also mobilized support from Saudi Arabia, which is widely believed to have matched the US dollar for dollar in support of the Afghan cause.[90] The Egyptians for some time provided training for the Afghan guerrilla fighters, and served as a major source of arms.[91] Equally important was the support rendered by China. But the country most essential to the success of the foreign military programme for the Mujahideen was Pakistan. Even under severe pressure and enormous difficulties, it did not budge from its stand on the Soviet withdrawal from Afghanistan. A broader coalition of forces, though expedient, in the end produced the result which even Ziaul Haq, the chief architect of the bleeding-the-Soviets policy, had earlier thought would be the 'miracle of the century'—the Soviet withdrawal.[92]

NOTES

1 Selig S Harrison, *In Afghanistan's Shadow: Baloch Nationalism and Soviet Temptations* (New York: Carnegie Endowment for International Peace, 1981), pp. 21-46.

2 According to Tahir Amin, about 600 Islamist workers were executed. See his *Afghanistan Crisis,* op. cit., p. 98.

3 *Anis* (Kabul), 21 July 1979.

4 For an opposite view *see,* Beverley Male, *Revolutionary Afghanistan: A Reappraisal* (London: Croom Helm, 1982).

5 Ashraf Ghani, 'Islam and Counter-revolutionary Movements', in John L Esposito (ed.), *Islam in Asia: Religion, Politics and Society* (New York: Oxford University Press, 1987), p. 82.

6 *See,* an excellent work by Olivier Roy, *Islam and Resistance in Afghanistan* (Cambridge: Cambridge University Press, 1986), pp. 54-83.

7 Author's interview with Pir Sayyid Ahmad Gailani in Islamabad on 23 August 1984.

8 Nakshbandia was founded by Muhammad Baha al-Din al-Bukhari Nakshband (1317-89). The Qadriyya order of dervishes is called after Abdul Kadir al-Djilani (d. 1166). The Chishtiya order is named after Muin al-Din Muhammad Chishti (1142-1236).

9 This is attributed to a prestigious school of Islamic theology at Deoband, India. The Deoband tradition, more or less like the Wahhabi movement, rejects the Sufi orders as being outside the pale of its puritanical Islam.

10 The Wahhabiya Islamic community was founded by Muhammad Abdul al-Wahhab (1703-87). The general thrust of the movement is to do away with all innovation which developed later than the third century of Islam. Wahhab's written polemic and that of his followers is almost entirely aimed at the cult of saints, as exhibited in the building of mausoleums, their veneration and visitation.

11 This was a frequent theme in the Marxist government propaganda against the resistance.

12 Syed Abul Ala Maududi, *First Principles of Islamic State,* translated and edited by Khurshid Ahmad (Lahore: Islamic Publications, 1974); *Islamic State: Political Writings of Maulana Syed Abul Ala Maududi,* compiled and translated by Mazharuddin Siddiqi (Karachi: Islamic Research Academy, 1986); Syed Abul Ala Maududi, *A Short History of the Revivalist Movement in Islam,* translated by al-Ashari (Lahore: Islamic Publications 1972).

13 Ibid.

14 On this *see,* several articles in *The American Journal of Islamic Social Sciences,* which is a biannual publication of the Association of Muslim Social Scientists and the International Institute of Islamic Thought, based at Herndon, Virginia.

15 Hassan Banna, *Five Tracts of Husan Al-Banna (1906-1949): A Selection from Mujmuat Rasail al-Guam al-Shahid Hasan-al-Banna,* translated from the

Arabic and annotated by Charles Wendell (Berkeley: University of California Press, 1978); Iftikhar Ahmad, *Ikhwan-ul-Muslimeen* (Faisalabad: Qirtas 1979).

16 Author's interview with Engineer Abdul Rahim (Jamiat), Media Information Department, Afghan Interim Government, held in Peshawar on 2 January 1990.

17 See *Foreign Report* (London) No.1675, 9 April 1981, p. 7.

18 Author's interviews with the Mujahideen leaders in Peshawar in December 1989.

19 Burhanuddin Rabbani's interview in the *Muslim*, 27 December 1987, *Supplement*, p. 2.

20 *New York Times*, 7 March 1990, p. A11.

21 *Far Eastern Economic Review*, 28 August 1981, p. 30.

22 'Shattering Mujahideen Unity' (editorial), *Nation* (Lahore), 31 July 1989.

23 *The Times* (London), 1 March 1990, p. 9.

24 *New York Times*, 26 May 1992.

25 *The Guardian*, 22 June 1980.

26 *New York Times*, 2 March 1980, section IV, p. 2.

27 Rahimullah Yusufzai's interview with Maulvi Yunis Khalis in the *Muslim*, 27 December 1987.

28 *See*, a report 'The Mujahideen Foreign Legion,' *Time*, 17 April 1989, p. 10.

29 Professor Sayyaf, *The Two Statements* (Place of publication and publisher n.a., 1980), p. 8.

30 Author's interview with Pir Sayyid Ahmad Gailani in Islamabad on 23 August 1984.

31 Author's interview with Dr Najibullah Mujaddedi, junior health minister of the Afghan Interim Government, on 27 December 1989, in Islamabad.

32 Kabul Radio in Pushtu, 7 May 1992, in FBIS-NES-090, 8 May 1992, p. 39.

33 Author's interview with Gailani in August 1984, in Islamabad.

34 This is at least the impression the author gathered during an interview with him, Ibid.

35 *Muslim*, 27 May 1992.

36 *Economist*, 2 February 1981, p. 52.

37 Author's interview with Dr Najibullah Mujaddedi. See note 13.

38 *Dawn*, 24 February 1989.

39 The Accord was signed in Peshawar, Pakistan, on 24 April 1992. See *Dawn*, 25 April 1992. On the peaceful transition of power from Mujaddedi to Rabbani, see *New York Times*, 29 June 1992.

40 *Sheikh* is a title given to religious scholars in Shiite Islam.

41 Ayatollah Mohammad Asif Mohseni's interview in the *Muslim*, 27 December 1987, Supplement, p. 2.

42 *Dawn*, 17 May 1992.

43 The following Shiite Parties agreed to unite under one command at the end of a conference convened by Iran:

1. Hizbullah
2. The Popular Front of Afghanistan
3. The Islamic Resurgence of Afghanistan

4. The Islamic Movement
5. The United Islamic Call
6. The Islamic Jihad Guards
7. The Islamic Forces Organization
8. Nasr Organization:
IRNA, Tehran, 23 December 1988, in FBIS Daily Report, Near East and South Asia, 29 December 1980, p. 56.

44 Author's interviews with Afghan resistance leaders in Peshawar and Islamabad in January 1990.

45 *See,* for instance, comments by Dr Farooq Azam, member of NIFA, in his 'Unity Among the Afghan Mujahideen,' *Central Asia,* No. 20, Summer 1987, p. 96.

46 *Dawn,* 2 January 1981.

47 *New York Times,* 2 June 1981.

48 *See,* the Charter of the United Islamic Front of Afghanistan which was circulated in 1983.

49 *See,* a report by a nearby observer, 'Afghanistan', *Bulletin of the Atomic Scientists,* Vol. 39, June-July 1983, pp. 16-23.

50 Author's interviews with resistance leaders in Islamabad and Peshawar in January 1990.

51 Maleeha Lodhi, 'Pakistan's ideas on interim government communicated to Cordovez', *Muslim,* 27 January 1988.

52 For this assessment *see,* Rosanne Klass, 'Afghanistan: The Accords', *Foreign Affairs,* Vol. 66, No. 5, Summer 1983, p. 936.

53 *See,* communique of the IUAM in *Afghan Jihad* (Peshawar), Vol. l, No. 3, January-March, 1988, pp. 26-27.

54 *Nation,* 7 January l987.

55 *See,* the text of the charter for Mujahideen transitional government in *Afghan Jihad,* Vol. l, No. 1, April-June 1988, pp. 22-6.

56 *Dawn,* 24 February 1989.

57 The ministerial positions at the time of election in February 1990 were as follows: Sibghatullah Mujaddedi (Afghan National Front) president and in-charge ministry of health, Abdul Rasul Sayyaf (Islamic Alliance of Afghanistan), prime minister and in-charge ministry of communications, Mohammad Nabi Muhammadi (Harkat-i-Islami), minister of finance and education and in-charge Supreme Court, Gulbadin Hikmatyar (Hizb-i-Islami), defence, scientific research and agriculture, Yunis Khalis (Hizb-i-Islami), interior, Haj and Auqaf, Burhanuddin Rabbani (Jamiat-i-Islami), foreign affairs, tribal affairs, and justice, Sayyid Ahmad Gailani (National Islamic Front), reconstruction, Islamic guidance, mines and industries. Commander Jalaluddin Haqqani was appointed chairman of the fourteen-member special committee.

58 *Dawn,* 27 February 1989.

59 *Dawn,* 21 February 1989.

60 *The Times,* 1 March 1990, p. 9.

61 Anthony Hyman, 'Winter of Deadlock in Afghanistan', *The Middle East,* November, 1981, p. 28.

62 Edward Girardet, 'Afghan guerrilla leader holds his own against Soviet offensive', *The Christian Science Monitor,* 2 October 1984.

63 Some of these were: commander Shafiullah in Ningrahar, commander Esmatullah Achakzai in Qandahar, commander Mohammad Shah Murtaza in Logar, commander Jalaluddin Haqqani in Paktia, commander Ismael Khan in Herat. In Hazarajat, ex-army Major Syed Hassan commanded the Shura's military operations.

64 Harvey J McGeorge II and Christine C Ketchman, 'Sabotage: A Strategic Tool for Guerrilla Forces', *World Affairs,* Vol. 46, No. 3, Winter 1983-4, p. 25.

65 *San Francisco Chronicle,* 18 June 1985, p. 13.

66 *The Christian Science Monitor,* 24 September 1981, p. 6.

67 *See,* a report by Barry Renfrew, 'Afghan Guerrillas Build Super-Base Underground', *News Press* (Santa Barbara), 8 December 1985, pp. 6-14.

68 Mark Urban, op. cit., pp. 190-4.

69 Darra is located in one of the tribal agencies, which have very little administrative control. The craftsmen of Darra have been manufacturing imitations of most of the best known light weapons since colonial days.

70 Aaron Karp, 'Blowpipes and Stingers in Afghanistan, One Year Later', *Armed Forces Journal International,* September 1987, p. 38.

71 The Mujahideen sources claim the kill-ratio of the Stinger to be close to the US test results, which appears inflated. *See,* a report '340 Stingers downed 269 Kabul aircraft', in *Frontier Post,* 8 July 1989.

72 Karp, op. cit., p. 40.

73 *The Nation,* 5 November 1986.

74 According to published reports the decision was taken in a meeting of top Pakistani civilian and military officials at which the US Ambassador was also present. The Director General of ISI was not in favor of attacking the city. Pressure came from the Foreign Office, which wanted to see the interim government established on Afghan territory. *See,* Henry Kamm, 'Pakistan Reportedly Ordered Rebel Attack on Afghan Town,' *International Herald Tribune,* 24 April 1989, p. 1; Mushahid Hussain, 'Afghanistan: After the Jalalabad Battle', *Frontier Post,* 7 April 1989.

75 Henry Kamm, 'For the Afghan Rebels, A Key Test of Discipline', *Jang* (Urdu) (Lahore), 23 September 1989.

76 *See,* a firsthand report by Abuzer Ghafari, 'Reasons for Delay in Mujahideen's Occupation of Jalalabad', *Nawai-i-waqt* (Urdu), 13 April 1989.

77 *See,* the comments of a leading scholar on this subject, Eqbal Ahmad, 'Two Battles, Two Wars', *Newsline* (Karachi), July 1989, p. 57.

78 *The Times,* 12 March 1990, p. 10.

79 *Washington Post,* 7 March 1990, p. 29.

80 Author's interview with Dr Faizullah Kakar, leader of Hizb-i-Islami, in Peshawar on 2 January 990.

81 *The Times,* 8 March 1990, p. 9.

82 Salamat Ali, 'Siege of Khost continues after Surrender Bid Fails', *Far Eastern Economic Review,* 26 April 1990, p. 20.

83 Professor Burhanuddin Rabbani reported that in the northern provinces, where his party had effective control, it would be difficult to equip even a regular fighting force of five thousand men. *Ukaz,* Jeddah, 31 March 1990, p. 3, in FBIS, 5 April 1990, p. 36.

84 For a fuller account of changes in American foreign policy, *see,* Fred Halliday, *The Making of the Second Cold War* (London: Verso, 1983).

85 'A Carter Doctrine' *Wall Street Journal,* 17 January 1980.

86 Richard H Shultz, Jr., 'Low-intensity Conflict: Future Challenges and Lessons from the Reagan Years', *Survival,* July/August 1989, p. 363.

87 Carl Bernstein, 'US Weapons for Afghanistan', *Chicago Tribune,* 22 July 1981.

88 Stephen J Solarz, 'When to Intervene', *Foreign Policy,* No. 63, Summer 1986, p. 26.

89 *Christian Science Monitor,* 13 March 1990, p. 7.

90 *Wall Street Journal,* 19 April 1984.

91 *Far Eastern Economic Review,* 2 October 1981, p. 9.

92 *Dawn,* 2 July 1987.

Chapter 8

Transition to Mujahideen Government

The collapse of the Marxist regime on 16 April generated domino effects, with the provincial and regional power arrangements undergoing an immediate transformation. Diverse coalitions hastily formed by Mujahideen parties, defecting military commanders and local tribal and ethnic leaders took control of towns and cities all over the country. This was, perhaps, the natural course in the circumstances, to fill the power vacuum left by the disintegration of the Marxist government. The Mujahideen parties based in Peshawar and the internal commanders confronted two difficult but central tasks: the formation of a broad-based transitional government and the reintegration of all the regional centres of power by bringing them under the authority of the central state.

The pursuit of these objectives, even in the normal political conditions of a democratic state, would be a complex process because it would require compromises, institutions, traditions, and some experience of pluralistic politics. Not only did the Mujahideen parties lack experience of coalition politics, but they had also been operating in a constitutional vacuum. In a political sense, the Mujahideen parties had not prepared themselves fully to take over power in Kabul. No prior agreements existed among them for sharing power, not because such efforts had not been made in the past under the prodding of foreign friends, but because each time an agreement was secured, it did not last very long. Political differences among the parties, fragmentation of political power in the country, and inexperience of democratic politics are the factors that have continually prevented the formation of stable coalitions among the Afghan parties. It is necessary to emphasize this point because the formation of any political coalition, and maintaining it, would require a democratic political framework.[1] Looking at the political realities of post-communist Afghanistan, it would be impossible to reunite the country or evolve a national consensus on the fundamental issues of restructuring the state and society without adhering to some democratic procedures. Since the appropriate institutions have yet to be created, and a value change has yet to occur, one might see a constant struggle between the hegemony and authoritarianism of one party, institution (military), or ethnic group, and the demand for access to power by all ethnic, religious and political groups on the basis of representation. In our view this struggle has already started to define the politics of Afghanistan, and its outcome will determine the future of the country itself.

Since the events in Kabul pre-empted the United Nations plan for a transitional government and a peaceful transfer of power, the second best option available at that time was an agreement among the resistance parties in Peshawar to form a coalition government for the interim period. The UN mediator Benon Sevan, Pakistani officials and the Mujahideen leaders had focused much of their attention on securing a wide consensus on the interim administration. The peace plan envisaged a fifteen-member council of neutral persons, to be selected from all the parties including Watan, to precede elections.[2] Najibullah was reported to have agreed to transfer all powers to this council and step aside but the sudden removal of Najibullah on 16 April made the UN-mediated plan irrelevant. It would be pertinent to mention here that Ahmad Shah Masud's decision to act in league with the Afghan army generals and the Uzbek militia leader Dostum quickened the fall of Najibullah. The question is, why did this coalition scuttle the UN plan? It may be that they did not expect a fair deal from the UN-sponsored political settlement, but the real motivation was political power. They took full advantage of Najibullah's vulnerable position, which had been weakened considerably by defections and revolts by the military commanders stationed in the north. Masud made a clever move by forging links with the disaffected commanders from the north. By ousting Najibullah through his allies in Kabul he both outflanked his Mujahideen rivals and succeeded in pre-empting the

Pushtun generals from capturing power. Here the question of who used whom is irrelevant, because the interests of the Uzbek-Tajik military commanders converged with those of the Mujahideen movement in the north. They seem to have had a common fear of Pushtun dominance. Had they not acted the way they did, the balance of political power might have tilted in favour of Hikmatyar and the Pushtun majority.

The Uzbek-Tajik coalition thus became better positioned militarily by moving first into Kabul in great numbers and occupying strategic positions Although Masud remained on the outskirts of the city, awaiting the formation of an interim government by the Mujahideen parties in Peshawar, for all practical purposes his decisions mattered the most in Kabul. His political posturing and military position also influenced the shaping of the Peshawar Accord. It was only after protracted negotiations and hard bargaining that, on 24 April 1992, ten of the Mujahideen parties reached agreement on a transitional ruling council.[3] The Peshawar Accord attempted to settle the issues of representation in the transitional administration and allocation of cabinet posts to various parties. It made the following provisions:

1. It created a fifty-member council, to be headed by Sibghatullah Mujaddedi for two months, to supervise the transition.
2. After two months, power would be transferred to Burhanuddin Rabbani, who would be the head of the state.
3. Mujaddedi would continue to be the head of the fifty-member council, which would acquire the status of a ruling council.
4. The transitional government, headed by a prime minister, would function for a period of four months, after which it would set up a new interim government.
5. The interim government would remain in office for a period of eighteen months, during which it would make preparations for elections and transfer power to the elected government.
6. The ruling council would comprise thirty field commanders, ten *ulema,* and ten members of the Mujahideen parties.
7. The prime minister would be nominated by the Hizb-e-Islami of Hikmatyar, and the first deputy prime minister by the National Islamic Front of Sayyid Ahmad Gailani, he would also be the foreign minister. The second deputy prime minister

would be appointed by the Hizb-e-Islami of Yunis Khalis, who would be education minister as well, and the third deputy prime minister would come from the Ittehad-i-Islami and would also hold the position of interior minister. Ahmad Shah Masud of the Jamiat-i-Islami was appointed as defence minister. The rest of the portfolios were to be divided among the remaining parties.[4]

The fundamental weaknesses of the transitional order were that it was structured in haste and was given legitimacy on the assumption that it would gain the support of other parties and groups controlling territories inside the country. It did not set any legal or institutional mechanism for placing the heavy weapons that the various groups still possessed under a unified command. Nor did the Accord make provision for settling any disputes among the parties which might arise in the course of implementation. How the interim government would bring under its control the independent fiefdoms that have sprung up all over the country is not specified in the Peshawar Accord. However, with all its shortcomings, the Peshawar Accord has provided a good starting point. In tackling some of these problems, the interim government will have to expand and even renegotiate its terms with the objective of integrating as many power groups in the country as possible. One has to appreciate the fact that chaotic political conditions, nurtured by fifteen years of civil strife, will require more than mere legal documents. What is needed is political accommodation and power-sharing among all the groups.

To be realistic, political and personal differences among the Mujahideen groups have persisted and may continue to prevent the necessary consensus on political arrangements. Their confrontations are symptomatic of group identities that have been strongly held and protected. With few exceptions, all the Mujahideen groups have fought independently. The rivalries that prevented political unity during the war of resistance have exacerbated the quest for power now that the common enemy has disappeared from the scene. Their stubborn divisions and the absence of democratic political institutions may further complicate the process of maintaining a stable coalition, even if broad agreement is achieved on basic issues.

One manifestation of these rivalries was the eruption of armed conflict between the forces of commander Ahmad Shah Masud and the guerrillas loyal to Gulbadin Hikmatyar over the control of Kabul following the collapse of the communist regime. An alliance of forces formed by Ahmad Shah Masud and Dostum evicted Hikmatyar's men from the defence ministry and presidential palace, alleging that he was conspiring to capture power with the support of KHAD, the notorious former intelligence agency, and Pushtun generals.[5] This led to further clashes on the periphery of Kabul. The reason is that Masud did not take into confidence any of the Peshawar-based parties, perhaps not even his own party leader, Rabbani, before entering into alliance with Dostum and commanders of the Afghan army.[6] Hundreds of Afghan lives were lost before common friends from Pakistan and Saudi Arabia, and some influential Mujahideen commanders, finally arranged a face-to-face meeting between Hikmatyar and Masud on 24 May that resulted in an agreement on a cease-fire, full implementation of the Peshawar Accord and acceleration of arrangements for holding elections.[7] Whether or not this agreement will hold or succeed in defusing their struggle for power is yet to be seen, but the bitterness between these two groups who have fought turf battles before may not disappear soon. Their conflict, if not resolved, may escalate into an ethnic war between the Pushtuns, fearing loss of power, and the coalition of Uzbek and Tajik groups from the north, who seem to have gained greater political influence in Kabul, which has been traditionally dominated by the Pushtun elite.

The politics of Afghanistan have grown too complex to be understood merely in terms of a rift between Hizb and Jamiat, extremists and moderates, or secularists and Islamic fundamentalists. The basic fact is that the country has fallen to many rival groups, along a pattern that was established by the independent territorial control of internal commanders. The territorial division is further complicated by the ethnic factor. The northern areas, with the exception of the Panjsher valley, underwent a tremendous transformation during the course of the war: they received greater development allocations from the Marxist regimes; the ethnic consciousness grew stronger; and the Uzbek Ismaeli community leaders organized militias in support of the Marxist regime. These militias fought bloody battles, and in many cases won, against mainly

Pushtun Mujahideen groups.[8] The northern ethnic elites also made tremendous gains in the hierarchy of the Marxist regime, both within the Party as well as in the governmental institutions. The transfer of power from the Khalq faction to Parcham, which began with the installation of Babrak Karmal in December 1979, facilitated their ascendancy to higher positions in the regime.

The conflict has changed the traditional social and political order of the country. The most important development is the emergence of ethnic identity. Non-Pushtun minority ethnic groups, such as Hazara, Tajik, Turkmen and Uzbek, have acquired a greater sense of empowerment by establishing autonomous centres of power. The authority of the traditional institutions that mediated social conflicts has declined. This has brought to the fore a new class of leaders, the resistance commanders who exercised political power in their respective areas of control quite independently of the former government in Kabul. These are some of the new political realities which the Pushtun majority and traditional elites have to recognize. The important thing is that political fragmentation of Afghanistan along ethnic lines presents a serious challenge to national unity.

This can be gauged from the pattern of realignments that followed the fall of Najibullah. It is widely known that Karmal's group in the Watan party pulled off the *coup* with the support of non-Pushtun military generals and Watan party leaders.[9] It could not be a coincidence that all the high-ranking generals and party members who were instrumental in removing Najibullah were Uzbek and Tajik, with a few Hazara Shiites.[10] It is more intriguing to note that they approached Masud, a Tajik resistance commander, assuring him of their co-operation in installing a Mujahideen government.[11] Ostensibly, this move was justified for averting bloodshed, but more importantly it marked a definite tilt of the Uzbek and Tajik generals, previously associated with the Marxist regime, toward Masud. It was logical and, perhaps, necessary for them, considering the collective interests of the northern provinces, which have acquired greater power in the course of war than was ever allowed to them by the dominant Pushtun elite.

Gulbadin Hikmatyar and a few other Pushtun resistance groups cried 'foul' over Masud's joining of hands with Dostum and other generals in forming the Jihad Council. Some of the Pushtun Mujahideen leaders have argued that continuation of the prominent generals of the Marxist regime in their previous positions was a negation of the spirit of *jihad*. The appointment of the Chief of Army Staff was a case in point. The president of the transition council first announced the appointment of Rahim Wardak, a former officer of the Afghan army who led the guerrillas of Sayyid Ahmad Gailani, but later cancelled it as a mistake, and instead authorized the continuation of General Muhammad Asef Dilawer in that position.[12] The making of such a key appointment would have required agreement among all the coalition partners, which has been difficult to obtain because of their divergent and conflicting interests. In opposing such decisions of the ruling council, the rival groups cite the hateful record of the Marxist regime's atrocities, co-operation with the former Soviet Union, and devastation of the country caused by their war. However, these objections lose significance in face of the fact that in 1990 Hikmatyar had supported Shahnawaz Tanai, who was then defence minister, in his bid to topple Najibullah. The Tajik-Uzbek coalition has also alleged that Hikmatyar's men sneaked into Kabul with the blessings of the Pushtun generals and Pushtun Watan Party functionaries, and that a number of Pushtun generals have defected to the Hizb-i-Islami.[13] Whatever the truth of these competing claims, allegations and counter-allegations, the new battle-lines may shape along ethnic lines. Only a decentralized political order that grants autonomy to all the ethnic groups can pave the way towards a genuine national reconciliation.

The presence in Kabul of the Uzbek militia controlled by Dostum emerged as both a political and an ethnic issue. As an ally, Ahmad Shah Masud has defended the role of this militia in Kabul and the new position of its leader as a full army general. In his view it was the defection of this force that paved the way towards toppling Najibullah; Dostum and his militia should be honoured rather than humiliated. Mujaddedi has praised him as a 'great Islamic warrior'.[14] On the other hand, Hikmatyar and a number of other Pushtun leaders have demanded the eviction of Dostum's militia, making it one of the conditions for supporting the interim government.[15] In their view, General Dostum's militia which was moved in, in advance by the

Tajik-Uzbek coalition, anticipating a similar march on the city by the forces of Hizb-i-Islami, proved too corrupt and disorganized to provide a credible or neutral security network. The very fact that the Uzbek militia was a partisan force, subjecting the rival Mujahideen guerrillas to its authority, did not win it any acceptance. An alternative has been suggested: a joint force of the Mujahideen parties. Unless this happens, or a neutral security force is organized strictly under the control of the new government, the Uzbek militia will be needed by the transitional council to defend the city against the forces loyal to Gulbadin Hikmatyar. Professor Rabbani recognized this problem more than did his predecessor, Mujaddedi. Within a week of assuming power, he ordered the disarming of all the armed groups roaming about the streets of Kabul.[16] He has strong feelings about the removal of the Uzbek militia from the capital.[17] It is quite evident that he would press for a disbanding of the militias, because removal of the Mujahideen guerrillas from the city while the militias maintain a strong presence would alienate most of his coalition partners.[18]

It is interesting to note that some of the Mujahideen groups based in the Pushtun areas and some prominent Pushtun leaders have echoed Hikmatyar's demands. Known Pushtun nationalists on the Pakistani side of the border, who never concealed their sympathy for the defunct Marxist regime, have objected to the composition of the transitional government, alleging that Pushtuns are being denied their legitimate share of power.[19] The Pushtun nationalists in the army, bureaucracy, and police loathe the prospect of Uzbek or Tajik control over Kabul.[20] As Hikmatyar commands the most formidable guerrilla force in Pushtun regions, he may emerge as a convenient choice for the nationalist elements in their struggle against the Uzbek-Tajik coalition, at least for the time being.

The main thrust of Hikmatyar's political strategy is not ethnic. However, he may appeal to this sentiment if Pushtun disillusionment grows further. He stands for cleansing the Mujahideen government of all those figures who were closely associated with the Marxist regime of Najibullah. He has argued that unless this is done, the real power would be wielded by the former communists, and the Mujahideen would remain 'hostages in their hands'.[21] Some of these fears have been confirmed by Mujaddedi. In one of his last press conferences before quitting the office of interim president in favour of Rabbani, he complained that many of his orders were not obeyed, that his defence minister did not allow him to go to the presidential offices, and that official radio and television consistently censored his statements.[22] This may be considered an outburst of frustration by Mujaddedi after failing to secure the support of Masud and Dostum for staying in power longer than justified by the Peshawar Accord,[23] but his evaluation of the balance of power in Kabul is not unrealistic. Much of the power has rested with former defence minister Ahmad Shah Masud, the Uzbek militia, and the remnants of the Afghan army.

It seems that three distinct groups are involved in the power struggle: Islamists, traditionalists, and the remnants of leftist-secular elements. Since the latter category stands discredited for its role in the Afghan tragedy, it may not reassert itself as an independent force for some time. For the time being, it would prefer to lie low while playing off one Mujahideen side against the other, thus acquiring some influence in the balancing act. By ideological orientation the leftist-secular elite would tilt towards the traditionalist parties and leaders if it fails to re-emerge as an independent political force. This inclination may gain strength if the Jamiat and Hizb succeed in forming a stable partnership. Given the ideological affinity between the two parties and their common historical roots, they may move closer to each other. This would depend largely on whether or not Masud reviews his alliance with the northern militia leaders. Once the reconciliation process between Jamiat and Hizb starts producing confidence, Masud might rethink the value of partnership with Dostum. Also, it would be necessary both to placate the fears of the Pushtun majority about northern dominance and to make some political inroads into the Pushtun regions. Professor Rabbani, who has worked in the Pushtun areas for a long time, would play a crucial role in diffusing ethnic polarization, and could even bring the Hizb and Jamiat leadership a bit closer to each other.

The nascent political system of the interim government closely resembles Shura, a system in which all the coalition partners would be consulted on vital policy issues. The leadership council provides the forum for debating all alternatives and options and serves as a mechanism for reaching consensus at the top. It might develop into an upper house of parliament in any future political restructuring. Another institution is the Jihad Council, which has about sixty-four members. Nearly all the Mujahideen groups, including Shiites and northern minorities, have been given representation on the Jihad Council. It is designed to act as the lower house of the legislature. In similar fashion, the offices of the president, prime minister, and cabinet ministers have been allocated to various parties in the light of the Peshawar Accord. But none of these councils seems to be truly functional. Also, neither the formal set of institutions nor the formal distribution of power explains where the real power rests. In the provinces, political power is more widely dispersed, while in Kabul Ahmad Shah Masud and his Uzbek-Tajik coalition enjoy far greater influence than any other party. The evolving system of Shura, which is an Afghan variant of western political coalitions, may work effectively if the real power at the centre is shared by all the partners. The process may confront political minefields, turbulence, and even paralysis because of the variety of actors involved in the power game and the interests of the external powers in its outcome.[24] The sectarian antagonism between the Wahhabi and Shiite groups has an explosive potential that may even rock the Mujahideen coalition. Their foreign connections would certainly aggravate tensions. It is no secret that Iran and Saudi Arabia have used their Shiite and Wahhabi clients respectively to play out their diplomatic rivalry in Afghanistan. This may further add to the difficulties of the interim government in reuniting the country [25]

The most immediate tasks before the new administration are how to re-establish the authority of the central government and how to disarm the heavily militarized society. The provincial centres are controlled by coalitions of various Mujahideen commanders, defecting military leaders and tribal chiefs. Their allegiance to the transitional government would depend on how the power-sharing arrangements among the rival political parties are worked out in Kabul. There are fears that if the interests of the aspiring guerrilla leaders and regional ethnic groups are not accommodated in the new power arrangements, they may form rival coalitions to oppose the new government or create their own autonomous administrations.

So far, the authority of the central state apparatus under the transitional council has remained weak. The shuras or coalitions which have captured provincial towns have been ruling independently. It will take time before the administrative structure of the state is revived and the writ of the central state is extended beyond Kabul. The very idea of a centralized state in the prevailing conditions of political fragmentation looks unrealistic—no ethnic, religious or regional group would like to give up its power to any arm of a centralized state. It would be both risky and impractical to reorder Afghanistan's future constitutional system along old lines. As the country has been carved up by a large number of groups of various categories, it would be pragmatic as well as necessary to devise a federal system with greater autonomy for the regions and provinces. This idea has yet to gain broader acceptance, although the demands of the ethnic and religious minorities are becoming more pronounced. Dostum, who controls much of the north-central and north-western provinces, has stated this demand categorically.[26] The leaders of Hizb-i-Wahdat have also put forward their concerns quite clearly, making their support of the transitional council conditional upon adequate representation and autonomy for their area.[27]

The alternatives to a federal system based on representation, autonomy, and meaningful power-sharing are chaos, anarchy and the frightening prospect of perpetual warfare in the name of national unity and integrity. Whether the Afghan leaders rise to the occasion and bring their country out of the present crisis as a unified entity is yet to be seen. Both the agony of the civil war and the political experience of their struggle suggest that the Mujahideen should also learn the rules of the game of politics: that sharing, not exclusive control, co-existence, not elimination of rivals, and democracy, not dictatorship would reunite their country.

The economic reconstruction and rebuilding of the country present another set of problems. Afghanistan has experienced one of the most brutal wars of contemporary history. The rehabilitation of refugees living in Iran and Pakistan and those dislocated within the country would require resources which are beyond the means of tragedy-stricken Afghanistan. With the ending of the cold war, leading to more inward-looking policies by the Western countries, and their greater interest in rebuilding the economies of the former Soviet Union and Eastern Europe, Afghanistan may not receive the amount of foreign aid that it needs to rebuild. A proper stock of the devastation caused by the former Soviet Union and its proxies in Afghanistan has yet to be taken. The costs of the environmental damage inflicted by sustained bombing of the fertile valleys, of the destruction of the fragile irrigation system, and of deforestation as a counter-insurgency measure, cannot be calculated in monetary terms. The scattering of millions of anti-personnel mines along the rural routes and in the fields poses obstacles to the restoration of normal human and animal habitation in the country

The best way to resolve the present crisis and end the state of political uncertainty would be to implement the Peshawar Accord in full and move faster towards preparations for holding elections. Procrastination or failure in doing so would raise afresh the question of political legitimacy. Making arrangements for an election in the current conditions, however, presents tremendous logistical and political difficulties, because one-third of the citizens are in exile,[28] and no administrative machinery exists to conduct fair and free elections. Besides the question of legitimacy, the fundamental issue is the creation of a national government capable of pulling the country out of the present chaos before it is too late to bring the fragmented society together. Against all odds and in the face of tremendous difficulties the Afghans have finally defeated the forces that caused the tragedy. Although they are divided and their rivalries run deep, they share a common interest in a unified and peaceful Afghanistan. Contrary to their portrayal as fanatics practising violence as the only means of settling their differences, Afghans have values, traditions and a political culture that may yet promote mutual accommodation. Although it is hard to predict the immediate structuring of a truly stable political order, a system of governance acceptable to the majority of the parties could start taking shape under arrangements that allow wider participation in power.

NOTES

1 Reference to a democratic framework should not be construed in western secular terms. In our view, democratic values evolve in cultural contexts and therefore should not be considered an end product. Also, democratic principles are not repugnant to Islam. As Islam defined the struggle against communism in Afghanistan, it would continue to influence the politics of the Mujahideen parties. It is equally important to mention that, although there are differences in the interpretation of Islam, it is the core value which is shared by all the parties.

2 Steve Coll, 'Fresh Hope Raised for Afghan Peace', *Washington Post,* 3 April 1992; *See also, New York Times,* 12 April 1992, p. 20.

3 The Accord was facilitated by Mr Nawaz Sharif, the Prime Minister of Pakistan. He used his leverage with the Pushtun parties to negotiate with other parties. A special Iranian envoy and UN Special Envoy Benon Sevan, besides several top Pakistani officials from the foreign office and the military, tried hard to bring the divided Mujahideen groups together.

4 *Dawn,* 25 April 1992.

5 These allegations were made by Dostum. See the report 'UN Envoy Calls on Rebels to Think as Afghans First', in *New York Times,* 22 April 1992, pp. 1,12.

6 This is demonstrated by the co-ordination between Masud and the Afghan army generals when he entered Kabul on 25 April. *See,* a report by Ahmed Rashid and Salamat Ali, 'Afghanistan: Peace With Pitfalls', *Far Eastern Economic Review,* 7 May 1992, p. 12. Pakistan's former Chief of Army Staff, Gen. (Retd.) Mirza Aslam Beg, who is one of the most informed former officials on the Afghan conflict by virtue of his top position in the Pakistan Army, and is now an articulate analyst of security issues, has argued that had Masud not planned to occupy Kabul in league with other generals, Hikmatyar would not have had to rush to Kabul from Peshawar. *See,* his statement in *Muslim,* 10 May 1992.

7 Among the common friends Ijazul Haq, son of the late Pakistani president Ziaul Haq, Saudi officials and commander Jalaluddin Haqqani played an important role in bringing Masud and Hikmatyar to the negotiating table. *See,* a report in *Dawn,* 26 May 1992.

8 Joseph Newman, Jr., 'The Future of Northern Afghanistan,' *Asian Survey,* Vol. 28, No. 2, July 1988, pp. 729-39.

9 *See,* a report by AFP in English, Hong Kong, 13 May 1992, in FBIS-NES, 14 May 1992, pp. 37-8.

10 Some of them were: Farid Mazdak, Abdul Wakeel, Sultan Ali Kheshtmand, Najmuddin Kaiwani, General Yar Muhammad, General Nabi Azimi, General Abdul Momin. *News,* 17 May 1992

11 Masud's understanding with them developed much earlier, perhaps in February or March when Afghan army commanders in the north started disobeying Najibullah's orders, which changed the balance of power. According to Masud they approached him, and he took advantage of the rift within the ruling group.

See, his interview in *Al-Haya* in Arabic, 7 May 1992, in FBIS-NES-093, 13 May 1992, pp. 45-6. Abdul Wakeel's meeting with Masud was reported in *New York Times,* 19 April 1992.

12 This incident also raised the important issue of who exercises the real decision-making power in Kabul — Mujaddedi, Ahmad Shah Masud the defence minister, or the remnant elements of the Marxist regime. Kabul Radio, Afghanistan Network in Pushtu, 7 May 1992 in FBIS-NES-92-090, 8 May 1992, p. 40.

13 Hong Kong, AFP in English, 7 May 1992, in FBIS-NES-92-090, 8 May 1992, p. 38.

14 *Dawn,* 24 May 1992.

15 Thousands of people demonstrated in Jalalabad, a Pushtun city in the east, against the presence of Uzbek militia in Kabul. Commander Abdul Qadir, head of the multi-party Shura in Jalalabad, rejected Masud's coalition with Dostum. Hong Kong, AFP in English, 10 May 1992, in FBIS-NES-92-091, 11 May 1992, p. 37

16 *Muslim,* 2 July 1992.

17 'A wise decision: implementation disappointing', (editorial), *Muslim,* 5 July 1992.

18 Since the disarming campaign was assigned to Dostum's militia and the forces of General Momin, the Mujahideen parties expressed grave concern. *Muslim,* 3 July 1992.

19 Begum Nasim Wali Khan, leader of the Awami National Party, accused Pakistan of trying to 'bypass Pushtuns and hand over power to non-Pushtuns'. Ajmal Khattak, Parliamentary leader of the ANP in the National Assembly, said, 'if it is a question of the honour of Pushtuns, then we will be compelled to support Gulbadin Hikmatyar' These statements were reported by Mushahid Hussain, 'An "Afghan solution" for peace,' *Nation,* 27 April 1992.

20 Ahmed Rashid, 'In Pashtun country', *Far Eastern Economic Review,* 14 May 1992, p. 12.

21 *See,* Hikmatyar's statement in *Dawn,* 22 June 1992.

22 *Dawn,* 25 June 1992.

23 Mujaddedi had argued that he was the only leader who had broad enough support to unite the country. He demanded to stay in office for at least two years. *New York Times,* 22 June 1992.

24 Edward A Gargan, 'The New Struggle for Afghanistan: On All Sides, Nations Seek Influence,' *New York Times,* 10 May 1992, p. 10.

25 In the first week of June, these groups fought bloody street battles in Kabul in which more than 100 persons were killed and many more injured. These clashes also affected the politics of the coalition. Mujaddedi allocated the ministry of interior to the Hizb-i-Wahdat, a Shiite party supported by Iran. Under the Peshawar Accord this ministry was assigned to the Ittehad-i-Islami of Sayyaf, which is a Wahhabi group supported by Saudi Arabia. See, *New York Times,* 7 June 1992, p. 21; M B Naqvi, 'Kabul: change of guard', *Dawn,* 4 July 1992.

26 Tehran, IRNA in English, 19 May 1992, in FBIS-NES-92-098, 20 May 1992, p. 44-5.

27 *See,* a statement by the spokesman of Hizb-i-Wahdat, Voice of the Islamic Republic of Iran (Persian), Tehran, 8 May 1992, in FBIS-NES-92-091, p. 35.

26 Tehran, IRNA in English, 19 May 1992, in FBIS-NES-92-098, 20 May 1992, p. 44-5.

27 *See,* a statement by the spokesman of Hizb-i-Wahdat, Voice of the Islamic Republic of Iran (Persian), Tehran, 8 May 1992, in FBIS-NES-92-091, p. 35.

28 According to press reports, four or five thousand refugees were returning to Afghanistan from Iran and Pakistan daily following the fall of Najibullah. *See,* a report by Barbara Crossette, 'US Helps Russians Search for Missing in Afghanistan', *New York Times,* 23 June 1992, p. 3. It was expected that a large number of them would return to their country towards the end of the year. This has not happened because of continued fighting between the forces of Prime Minister Gulbadin Hikmatyar and President Burhanuddin Rabbani. *New York Times,* 30 December 1993, p. 4.

Chapter 9

The Vortex of Regional Power Rivalry

Afghanistan's location at the junction of three strategic regions—South, South-West, and Central Asia—increases its importance for its neighbours and also makes it vulnerable to their adverse influences. Since the military intervention of the former Soviet Union, the factors of regional and international power politics have added an important dimension to the internal politics of Afghanistan. Much of this has been manifested in the client-patron relationship between the Afghan parties and external powers. The roots of external intrusion into Afghanistan lie essentially in the internal confrontations. The quest for power, control of territory and eventual political legitimacy compelled the Afghans to seek foreign support and friends—but certainly at the loss of some freedom to act independently. This was true of the Marxist regime that invited the Soviet troops to help stabilize itself, and also of the groups opposed to it for reasons of ideology and its subservience to Moscow. Whatever the rationalization of foreign connections, they are never completely divorced from the interests of the suppliers of weapons and money who are no more altruistic than those involved in the struggle for power in Afghanistan. Among the regional states, Iran and Pakistan have been affected by the Afghan conflict, and in return have influenced the politics of the resistance against the communist rule. What were the political and security interests which shaped the Afghan policy of the regional states? What role did they play in strengthening the Afghan resistance against the former Soviet Union? How would they attempt to play out their rivalries in Afghanistan? What will be the impact of the independence of the Central Asian states and the fall of the Marxist regime in Afghanistan on regional security? It is to these questions, and the stakes the of regional states in developments within and without Afghanistan, that we now turn.

PAKISTAN

Pakistan played a critical role not only in mobilizing international support for the Afghan Mujahideen but also in equipping and training their forces. Given the complex economic, political and security problems confronted by Pakistan, this role was not an ordinary foreign-policy response to a crisis in the region. Why was Pakistan willing to take risks to complicate the situation for the Soviet Union by playing a key role in strengthening the Afghan resistance? There are two explanations, one with reference to the domestic politics of Pakistan, and the second related to the widely-circulated perceptions of geopolitical destabilization in the region caused by the Soviet military action against Afghanistan.

There is no denying the fact that the military regime of General Ziaul Haq was isolated after he executed former Prime Minister Zulfiqar Ali Bhutto in April 1979. Even before that the martial law government was shaky, confronting questions of legitimacy and demands for the fresh elections which it had promised to hold within ninety days when it toppled the civilian regime. The Soviet military intervention in Afghanistan coincided with the revolution in Iran. These events, though not inter-connected, undoubtedly constituted a major change in the security relations of the south-west Asian region. The demise of the monarchy in Iran, in particular, knocked the cornerstone out of the United States' strategic edifice in the Gulf. Pakistan's willingness to play a front-line-state role in the matter of Afghanistan was thought to compensate somewhat for the loss of Iran in the US endeavours to restructure a new strategic consensus. By promoting a convergence of interests with the United States on the Afghanistan issue, General Zia benefitted in two ways. First, he strengthened his support base in the civil and military bureaucracy by bargaining for US economic and military assistance. Second, his strong support for the Afghan resistance placated some of the right-wing religious and political parties and influential groups within the power-structure of the state. The economic component of the US aid package, along with the rescheduling of World Bank loans and a greater flow of resources from other countries of the western bloc, injected new life into the Pakistani economy. The resultant economic growth was a critical factor in hedging the Zia regime against the democratic forces in the country.

While the political motives of the regime were a strong factor in dictating a militant policy, the strategic consequences of the Soviet occupation of a non-aligned Muslim country that had historically played a buffer role could not be ignored. Even if Islamabad's posture had been less confrontational, it could not have escaped the spill-over security effects of the Soviet war in Afghanistan. Within months of the Soviet intervention, the border regions of Pakistan were over-run by hundreds of thousands of refugee families, and the number soon swelled to millions. Given the historically 'free' character of the tribal areas and the porous borders, Pakistan was in no position either to stop the inflow of refugees or to effectively control the movement of the guerrillas from and into Afghanistan. Moreover, in the past, successive Afghan governments had menaced Pakistan's security by giving material and political support to the separatist elements in its Balochistan and Frontier provinces. The active military involvement of a superpower in support of a narrow-based regime increasingly hostile to Pakistan presented difficult security problems. At the time Pakistan's Afghan policy was taking shape, the military leadership drew up the following scenarios of contingencies presented by the Soviet military presence in Afghanistan:

1. Kabul-Soviet attacks on the refugee camps, and hot pursuit of the Afghan guerrillas into Pakistan territory.
2. Soviet military pressures along the Durand Line with the aim of seizing salients of Pakistani territory in order to control mountain passes.
3. Encouragement and support of ethno-national movements inside Pakistan.
4. Either an Indian attack on the eastern border or a co-ordinated attack, with the Afghan-Soviet forces opening a western frontier.[1]

None of these fears was realised, but this type of strategic thinking deeply influenced the Pakistani perceptions of security. One way

Pakistan thought it could avert the dangers from Afghanistan was to make the Soviet goals—of stabilizing the proxy regime in Kabul and pacifying the country—difficult. At times this policy appeared unrealistic, perhaps provoking Soviet retaliation, but it seems Pakistan's assessment of Moscow's vulnerabilities was correct. As the uprising in Afghanistan became nation-wide and the resistance began to deny the Kabul-Soviet forces control over the countryside, Pakistan's policy of increasing Soviet expenditure was reaffirmed.

The most important factor in strengthening Pakistan's resolve to upgrade the capacity of the Afghan resistance was the security commitment and military assistance it received from the United States.[2] What motivated Washington to support the Afghan resistance and Pakistan was its strategic interest in preventing the Soviet Union from further expanding its political and military influence to the south-west Asian region. Without adequate political and military support, Pakistan might have capitulated to Soviet pressures and accepted what it was told to accept as 'political reality.'

On its own, Islamabad had limited capacity to defend itself against a determined Soviet invasion, but Pakistan succeeded in counter-balancing the Soviet threat from Afghanistan by a variety of means. First, since the Soviet invasion of Afghanistan raised the spectre of geopolitical destabilization in the adjacent region of south-west Asia, Pakistan emerged as central to the US-sponsored strategic consensus in the region.[3] This convergence of interests with Washington enabled Islamabad to obtain economic and military assistance to modernize Pakistan's armed forces, which were in a dismal state.[4] Pakistan's security was further enhanced by the emergence of a broad coalition of international forces supporting Islamabad's effort to expel Soviet forces from Afghanistan.

Second, Pakistan collaborated, at considerable risk, with the Afghan resistance to develop a guerrilla-war strategy that thwarted Moscow's attempts to consolidate its hold on Afghanistan and to pacify the country. Benefiting from sanctuaries in Pakistan and from the aid funnelled through Pakistan, the Afghan resistance pinned down Soviet power, forcing Moscow to fight a reactive counter-insurgency. The ensuing stalemate not only tarnished the Soviet image of invincibility but ultimately raised to an unacceptable level the cost of Moscow's protracted engagement. Soviet preoccupation with a war of attrition, resentment at home, and the opprobrium leveled at the USSR internationally prevented Moscow from carrying out its military threat against Pakistan.

Finally, Pakistan was careful not to escalate the level of resistance to a point that would provoke Soviet retaliation. Pakistan did this mainly by controlling the technological level of weapons supplied to the resistance while also judging well Soviet vulnerabilities, turning a blind eye to the supply of high-performance weapons to the Mujahideen at critical phases of the conflict. On the diplomatic front, Pakistan attempted to placate the Soviet Union by responding favourably to the proposal of a political settlement arranged through the auspices of the United Nations, while remaining firm in its position that Soviet withdrawal from Afghanistan was a prerequisite to such a settlement and that armed conflict would continue until such withdrawal occurred.

Although the termination in February 1989 of direct Soviet military involvement in Afghanistan removed the major threat to Pakistan, the unfinished war continued to occupy a central place on Islamabad's diplomatic agenda. New streams of Afghan refugees trickled into Pakistan as the internal conflict intensified after the Soviet withdrawal. As long as the Najibullah regime remained in power, Pakistan continued to support its removal through the Afghan resistance, and continued to bear the burden of refugees.

Beyond the economic cost of feeding the refugees and the ecological disaster caused by their encampment in certain areas, their continued presence in Pakistan is a potential threat to social tranquility and political stability in the country.[5] After the formation of a Mujahideen government in Kabul, about 4,000 to 5,000 refugees were reported to be returning daily to their country from exile in Pakistan and Iran[6], but a large number will stay back until economic and political conditions improve Most of the refugees are politically connected with the various competing Afghan resistance parties and have ready access to arms. Pakistan itself is a polarized society, and given the the current state of confrontations both political and sectarian, wider social conflagration cannot be ruled out. Over the past decade, national boundaries among the Afghan and Pakistani political groups have blurred. The extension of political and material support to groups sharing a particular political vision, or even acting in self-interest but pursuing parallel goals across the border, has become an accepted norm. Thus, it is virtually impossible to isolate Pakistan's domestic political conflict from the Afghan factor, at least until a broad-based government in Afghanistan is formed and the refugees are peacefully repatriated.

The decision of the government of Prime Minister Nawaz Sharif to give its full support to the UN peace plan, and then to help form the Interim Commission under Sibghatullah Mujaddedi, did not accord with the self-defined political interests of the more militant Mujahideen groups that Islamabad had for a decade favoured over more moderate forces. Some of Pakistan's own groups with considerable potential for agitational politics, e.g. the Jamat-i-Islami (Islamic Party), have reacted sharply to this shift in the government's Afghan policy. In protest, the Jamat-i-Islami withdrew from Pakistan's ruling Islamic Democratic Alliance (IJI) in May 1992. Growing resentment might put various combinations of Afghan and Pakistani political forces on a collision course with Islamabad. This is a worst-case scenario, but it reminds us that if the Afghan crisis is not settled by the creation of a Mujahideen government enjoying the broad support of all important Afghan resistance factions, there could be serious consequences for Pakistan's politics and security.

In confronting the present crisis in Afghanistan, Pakistan has to reckon with a falling-off of the wide-ranging international support it enjoyed before the collapse of communism. In the eyes of the West, the threat of geopolitical destabilization is past and the Afghan conflict is widely perceived to have degenerated into a civil war. The rapid transformation of Moscow-Washington relations from rivalry to co-operation has already started affecting Islamabad's regional priorities, not to mention the outcome of the civil war in Afghanistan.

In any case, the factors that brought Pakistan to the forefront in US strategic planning in the area seem to be evaporating. Two developments in particular initially raised Pakistan's stock in the US calculations: (1) Soviet intervention in Afghanistan, and (2) the Islamic revolution in Iran. One can discern a qualitative change in the regional environment on both counts. The concerted response of Washington, Islamabad, and the Afghan resistance did finally convince the Soviet Union that the war in Afghanistan was not winnable and that it could not achieve any success in restructuring relations with the West without extricating itself from such regional conflicts. Likewise, fears that Iran's revolution might destabilize the geopolitics of the Gulf region have significantly receded as Tehran continues to reel from the devastation experienced in its war with Iraq. Moreover, the balance of power in the region has drastically shifted in favour of the US and its Western allies since the destruction of the Iraqi armed forces. The US military presence in the region and the very successful use of force against Iraq should discourage, for some time, any local actor from threatening Washington's interests.

As the fundamental shift in Soviet policy towards Afghanistan removed the most compelling reason for US economic and military assistance to Pakistan, the issue of nuclear proliferation returned to the dominant place it once occupied on the US-Pakistan bilateral agenda. Even the military assistance provided to Pakistan had been designed in part to afford Washington a degree of influence over Pakistan's nuclear programme, and modernization of Pakistan's armed forces was meant to reduce the security incentives for Islamabad to exercise the nuclear option. For their part, based on a consensus among a wide array of political opinion within the country, various Pakistani regimes, both military as well as civilian, have linked their acceptance of total international safeguards to what India does with its own nuclear programme.[7] The United States has grown increasingly suspicious about Pakistan's nuclear programme, charging Islamabad with continuing to produce fissionable material in violation of pledges made to qualify for US economic and military assistance. Reflecting changed global and regional priorities, Washington brought its security relationship with Pakistan to an abrupt end in 1990, when President George Bush refused to issue to the US Congress the required annual certification that Pakistan's nuclear programme was peaceful.[8]

With the Afghan war apparently stalemated and with international support dwindling, Pakistan found itself in an increasingly difficult situation. It faced two hostile neighbours—India, and Afghanistan under the communists—which had converging interests in neutralizing the Pakistan factor in the Afghan conflict. Acting together, they could exacerbate ethnic conflict within Pakistan. Moreover, India had enormous military power with which it could increase pressures on the Pakistani borders.

To these negative incentives impelling Pakistan toward more active promotion of a political settlement in Afghanistan, there has been added a positive incentive—the prospect of productive political and economic relations with the newly independent republics of former Soviet Central Asia. Islamabad sees enormous opportunities for entering the markets of these states, which are equally eager to establish trade and investment links.[9] But without the bridge of Afghanistan, Pakistan may not succeed in furthering its interest in this region. Moreover, the present political leadership of the Central Asian states is opposed to the militant Islamist Afghan groups that were long favoured by Pakistan. Many of these leaders are recent converts from the Communist Party and are committed to a secular approach to modernizing their Muslim societies. Having observed the turmoil next door in Afghanistan, they are genuinely concerned about the spill-over effects that a triumph of radical Islamist forces there might have. They fear that if fundamentalists of the Pushtun ethnic group were to capture power in Kabul, the Uzbek, Tajik, and Turkmen ethnic minorities of Afghanistan might feel insecure and start migrating to Central Asia. Pakistan has been told in unequivocal terms that were it to support the establishment of an extreme right-wing government in Afghanistan, that would impede the growth of friendly relations with Central Asia.[10]

For all these reasons, Pakistan moved to promote more actively the formation of a broad-based transitional government dominated by the Mujahideen as a means of achieving national reconciliation in Afghanistan. Pakistan's policy shift reflected a growing consensus on the Afghan issue among all elements of the Pakistani power structure (president, prime minister and the military), a consensus even shared with the political opposition then led by Benazir Bhutto.[11] In this respect, Prime Minister Nawaz Sharif and his associates appeared to be more pragmatic than their predecessors, and committed more to tackling pressing economic and social issues at home than to running after the mirage of military victory on the part of its Afghan Mujahideen clients. The military establishment of Pakistan itself drifted away from its previous position after re-evaluating the capacity of the Afghan resistance groups to force a military solution.[12] Realizing the weakness of the Mujahideen, the military is understood to have given its blessings to the diplomatic initiatives of the Pakistan Foreign Office.

The collapse of the Marxist regime in Kabul a mere 105 days after the Russian aid cut-off dramatically transformed Pakistan's security environment. The installation of an interim Mujahideen government represented at least partial achievement of the decades-long quest by Islamabad for a friendly Afghanistan. The possibility of a Russian or Indian exercise of any influence in this strategic back yard of Pakistan was essentially eliminated.

This does not mean that Pakistan's relations with Afghanistan will necessarily be smooth, without irritants. Pakistan now faces the dilemma of how to react to the ongoing power struggle among Afghanistan's new leaders. Active involvement to promote reconciliation may be branded as undue interference, while indifference to the infighting may cause further fragmentation. The moderate groups now in power, as well as such coalition partners as the Uzbek militia and commander Masud, have long resented Pakistan's support of the Hizb-i-Islami, and even now see in Hikmatyar's opposition to the interim government the hand of Pakistan's Inter-Services Intelligence. At the same time the Islamists accuse Pakistan of sacrificing their interests by supporting the coalition headed by Rabbani.

The ethnic issue also complicates Pakistan's policy on Afghanistan. The current Mujahideen government is dominated by non-Pushtun officers and resistance commanders. This is partly the reflection of social and political changes forced by the long civil war: for the first time, ethnic and religious minorities have acquired a real sense of empowerment. In the course of the war, the Shiites of Hazarajat in central Afghanistan and the Tajiks in the northeast under Masud established effective administrative structures in their respective liberated zones. Having acquired autonomy and exercised political power, they may never again accept domination by the Pushtun majority. For their part,the Pushtuns tend to view the way in which the Marxist regime was overthrown as a conspiratorial power play by non-Pushtun army officers in collaboration with Masud. Given this political fragmentation along ethnic lines, it may be that only a decentralized political order granting autonomy to the ethnic groups can pave the way towards a genuine national reconciliation.

CENTRAL ASIAN STATES

As suggested above in the context of the Pakistani calculation, the independence of the Central Asian Republics of the former Soviet Union has added a new factor to the vortex of regional power rivalries in which Afghanistan finds itself today. Tajikistan, Uzbekistan and Turkmenistan share not only borders but also an ethnic, cultural, and religious affinity with Afghanistan and its peoples.

The elites in Central Asia perceive Islamist movements in the tri-junction of South, South-West, and Central Asia as highly destabilizing. These elites—by and large still the product of the Soviet communist system—are groping for a political identity, for a broader political base. Their adherence to the authoritarian power structures of the old political order continues but they face growing opposition on the issues of democratic liberties and greater religious freedom. The growing power of the Islamist forces and their alliance with the secular democratic parties were visible in the brief success they achieved when they removed the ruling establishment of Tajikistan in May 1992.[13] The Central Asian elites feel their legitimacy threatened by cross-border, transnational Islamic influences emanating from Afghanistan, although the current Mujahideen government in Kabul has assured them it does not intend to align itself with the opposition forces in the republics.[14]

The role of the Central Asian states may acquire greater significance if the ethnic split between Afghanistan's Uzbek and Tajik groups on the one hand, and the Pushtuns on the other, becomes unmanageable. Ethnic bonds, with family and blood ties, may push the Uzbek and Tajik minorities of Afghanistan towards their co-nationals across the border in the event that the current power struggle in Kabul results in a resurgence of the Pushtun element. So far, the Central Asian states have played no role at all. This is probably due more to the sorry state of their economies and the current crisis of state formation than to any lack of interest in Afghanistan.

Given the facts that road communication between Uzbekistan and northern Afghanistan is quite well developed and that there are thus considerable opportunities for trade, the Central Asian states can be expected to pay increasing attention to political events in Afghanistan. With the barriers previously erected by Moscow now fallen, one might even see the transformation of the entire region of Central Asia, Afghanistan, Iran, and Pakistan into a zone of economic and political cooperation. Azerbaijan, Turkmenistan, and Tajikistan have already joined the Economic Co-operation Organization formed by Iran, Pakistan, and Turkey. Two similar but competing regional oraganizations—the Black Sea and Caspian Sea co-operation zones—initiated respectively by Turkey and Iran, have attracted the Central Asian states. A summit conference of the heads of states and governments of the Central Asian states along with Iran, Pakistan, and Turkey in Ashkabad in May 1992 was another milestone.[15] A number of factors have forced the Central Asian states to look in the direction of the Islamic neighbours. First is the disruption of traditional economic ties with Russia and the other European republics of the former USSR. These republics supplied almost all the manufactured goods and fuels to Central Asia in exchange for raw materials.[16] With economic chaos and instability, they are no longer a reliable source. Second, Russia is reorienting itself entirely towards western Europe, leaving Central Asia to fend for itself. Third, the bulk of Western aid that has been committed to the Commonwealth of Independent States (CIS) has been channelled into the European member-states, mainly aimed at Russia.

But this does not suggest that final choices have been made. The influence of Russified elites, the Russian language, and the settlement of the ethnic Russian population would complicate any quest to forge historic and cultural ties towards the East. Secondly, the Central

Asian states might emerge as a new arena of rivalry and competition among the regional states. Iran and Turkey represent two extremes of the political spectrum and strategic alignments, while Pakistan, another competitor, blends a moderate Islamic outlook with a democratic system. A consensus on economic and political forms and on state identity has yet to emerge in any of these states, but the need for economic reconstruction in Afghanistan and Central Asia in line with regional and global trends towards marketization and liberalization may not only bring these countries closer to one another but also inject some cohesiveness into their policies.

INDIA

India's Afghan policy during the period of the Soviet interventionary war was contradictory. Although New Delhi vociferously decries the involvement of external powers in the region's security affairs, it maintained a guarded silence over the Soviet invasion of Afghanistan. Officially, it never uttered a word against the atrocities and devastation caused by the Soviet forces. Such silence sharply contrasted with the indignation unequivocally expressed by most Third World countries over the Soviet actions.

India's pro-Soviet position on the Afghan issue was, however, consistent with New Delhi's general line on global alignments. First, it did not see any point in offending Moscow over a situation that did not directly endanger Indian security. Second, India appeared more worried about the implications for the regional balance of power should the Mujahideen forces topple the PDPA regime and install a government friendly rather than hostile towards Pakistan; what was desired was a secular, neutral, and friendly Afghanistan.[17] Finally, India thought that Pakistan, drawn into the Soviet-supported conflict, would be amenable to New Delhi's leadership role, or at least would emerge still weaker in the regional power equation.

Given these calculations, India adopted a policy that was supportive of the Soviet version of events in Afghanistan.[18] It was no less vocal than the Soviet Union in articulating the view that Pakistan, through its support of the resistance, had attempted to destabilize the situation inside Afghanistan. On this ground, it tended to interpret the Soviet intervention as reactive rather than strategically offensive. Such a characterization seemed designed to undermine any justification for US military assistance to Pakistan.[19]

However, New Delhi misjudged the magnitude of Moscow's offence and the extent to which Pakistan was willing to confront the Soviet Union on the Afghan issue. By downplaying the Soviet aggression, India found itself neutralized in a major conflict next door. Pakistan, on the other hand, emerged a far more influential regional actor, winning considerable international support.

Unlike its neighbours and the overwhelming majority of Third World countries, India never questioned the political legitimacy of the various Afghan Marxist leaders installed by Moscow, from Nur Muhammad Taraki in 1978 right down to Najibullah.[20] Rather, it continued business as usual, extending technical and economic assistance to the beleaguered regimes.[21] Indian interest in the fate of the Kabul regime grew even stronger as it became clear that the Soviet Union was serious in ending its Afghan war. New Delhi first attempted to be part of a regional settlement, but this was frustrated by Pakistan and its allies. Having been marginalized on the Afghan diplomatic scene, India increased its support for the Najibullah regime.[22] India calculated that exhaustion from the war would force the Mujahideen opposition in the end to accept Najibullah's rule. This perception proved wrong: it was Najibullah who finally caved in under pressure from the Afghan resistance and his regime's own internal divisions.

By opposing the cause of national self-determination, India alienated itself from the popular forces of Afghanistan. Although many of the Afghan opposition leaders could appreciate the closeness of India's ties with Moscow, they expected at least a neutral position regarding the war in Afghanistan.

New Delhi's approach overlooked the probability that the future politics of Afghanistan would be more complex than the simple alternatives between secularism and fundamentalist Islamism. In the end, the Afghanistan that may emerge out of the present chaos may be neither controlled by Pakistan nor representative of Islamic fundamentalism. Rather, the country's traditional Islamic character and true independence could conceivably afford much room for co-operative relations with both India and other counties in the region. Unfortunately, India's unflagging support for the Marxist regime in Kabul may preclude New Delhi's drawing benefit from this result. The Mujahideen government has indicated that Kabul will have to review its relations with New Delhi in the light of the latter's persistent opposition to the struggle of the Mujahideen against the Soviet Union and its Afghan clients.[23]

Antagonistic role images inherent in the psyche of national power structures have consistently prevented the defusing of Indo-Pakistani tensions. Attempts to delink confidence-building measures from the contentious issues until such time as the severity of disputes is significantly reduced, or enough goodwill is created to resolve them, have achieved only marginal success.[24] Hangovers from the past have, time and again, caused a return to adversarial relations. Kashmir is a case in point.

The fall of the Marxist regime and the formation of the Mujahideen government in Kabul may influence the political and psychological dynamics of the ongoing conflict in Kashmir, over which India and Pakistan have fought three costly wars. Long before capturing power in Kabul, the Mujahideen leaders supported the liberation of Kashmir, occupied by Indian forces in 1948 after an instrument of accession was signed by the ruler of the state, obviously against the will of his mainly Muslim subjects. Although India pledged itself to conduct a plebiscite in Kashmir to determine the final status of the territory,[25] it has yet to be held, deepening the suspicion and hostility in relations with Pakistan.

Since the Simla Agreement of 1972 with Indian, Pakistan has preferred to set the issue on the back burner, and has only kept it alive symbolically for domestic consumption[26], but with the eruption in 1989 of a mass movement for self-determination in Indian-occupied Kashmir, the issue has once again moved to the fore of the India-Pakistan agenda.[27] Unlike in similar situations in the past, both countries seem to be showing considerable restraint. Nonetheless, the two sides have regularly exchanged accusations over the past years, producing a war hysteria.[28] The current uprising in Kashmir is so intense and widespread that India may not be able to control the situation effectively, even if Pakistan is intimidated into restricting itself to a passive diplomatic role.

Islamabad has, for historical reasons, remained closely interested in the political struggle of the Kashmiris, but it sees the Kashmir uprising against the Indian occupation as a legitimate movement that has developed indigenously.[29] The current uprising stems primarily from the alienation of the Kashmiris from the Indian state structure, constant manipulations of Kashmiri political groups by the New Delhi authorities, and a strong sense of nationalism which never died during the past four decades. Indiscriminate acts of violence committed by the Indian security forces have further worsened the situation.

There is genuine fear in Pakistan that if India's current repressive measures in Kashmir fail, New Delhi may resort to war, citing Pakistan's involvement in Kashmir. India's decisive military edge over Pakistan and its ability to sustain war for a longer period of time may tempt some of the hawkish Indian leaders to 'resolve' the issue by attacking Pakistan with the object of destroying its military capabilities. As long as Soviet involvement in Afghanistan continued, New Delhi felt constrained from pressuring Pakistan for fear of triggering increased flows of Western military assistance to Islamabad. So far, neither country appears to desire yet another costly war, one perhaps more destructive than the previous ones, given the high-technology weaponry the two sides possess. Two factors, however, may aggravate tensions. The first is public sentiment, which has run far hotter than the relatively cautious statements of government leaders. The second is the upsurge of fundamentalist Hindu elements since the 1991 Indian elections. While conservative and overtly anti-Pakistan politicians still sit in the opposition, 'Crush Pakistan' may yet emerge as a popular political slogan in an India wracked by a mood of uncertainty. The cruel fact is that the decisions that would affect the future security of the region may not be completely divorced from the domestic political compulsions of both India and Pakistan.

IRAN

Although Iran has stated that unity among all the Afghan resistance parties is necessary if there is to be an effective political alternative for Afghanistan,[30] Tehran's support has remained mostly confined to the Shiite factions. After its disengagement from the war with Iraq, Tehran stepped up aid to its Afghan clients and paid more attention to their politics, although this involvement never grew to a level that might have been decisive in any competition between Shiite resistance groups. Their efforts were crowned with remarkable success in March 1990, when eight Shiite parties met together for the first time, in Tehran, to develop a common political and military strategy, and formed an alliance under the Hizb-i-Wahdat (Unity Party).[31] A related objective had been to seek greater representation of the Shiite parties within the resistance as a whole and in the Mujahideen successor regime in Kabul. Lack of agreement on Shiite representation prevented those parties from joining the Pakistan-backed interim government-in-exile in 1990.[32] The Pakistan-based parties reportedly refused to accept twenty-five per cent representation for the Shiite groups on the Afghan Shura (the consultative assembly that elected the interim government), contending that this figure overstated the proportion of the Shiite population in Afghanistan.[33]

Tehran has been ambivalent towards the Pakistan-based parties for at least two reasons. In addition to its discomfort with the traditional, non-revolutionary Islam of the majority of the Sunni parties, Iran never felt comfortable with the flow of US support to such Afghan resistance parties. *Realpolitik* considerations also play a part, as in Tehran's opposition to parties such as the Hizb-i-Islami of Gulbadin Hikmatyar. Deprived of Soviet support, the Najibullah regime played upon Tehran's concerns, indicating an apparent willingness to grant autonomy to the Shiite parties in return for their support for an independent deal for Kabul. In exchange, Iran facilitated Najibullah's November 1990 dash to Meshhad, where he conferred with some of the Mujahideen leaders, including Pakistan-based moderate Afghans.[34] Although this was an indication that Iran might negotiate with Kabul if the interests of the Shiite parties were ignored by the Pakistan-based resistance groups, there was no evidence that Iran had reversed its policy of co-operation with Pakistan in seeking a settlement of the Afghan issue.

Iran has been an important backstage player in the formation of the current Mujahideen coalition government in Kabul.[35] Tehran expects to obtain a better deal for the Shiite groups it has supported and greater influence for itself by patronizing moderate Sunni parties and the Tajik and Uzbek Mujahideen commanders. Iran continues to co-operate with Islamabad in trying to work out a political settlement in Afghanistan. The two co-operated in convening two trilateral conferences (Iran, Pakistan, and Afghan resistance groups) that brought Sunni and Shiite groups together on the same platform. Ultimately, though, it will be up to the Afghans to determine whether a suitably representative political authority can be created in Kabul that will satisfy the diverse Mujahideen groups. In case of the fragmentation of Afghanistan into smaller units, Tehran might be tempted to offer greater support to Shiite groups in Hazarajat and Hirat regions.

An improvement in Iran-Saudi relations may also have some impact on the Afghanistan situation. Rivalry between Tehran and Riyadh has been muted by common antipathy to Iraq's Saddam Hussain. The two countries have restored diplomatic relations, which were severed during the Iran-Iraq war. A triangular consensus involving Iran, Pakistan, and Saudi Arabia has finally developed in support of the accord reached in Peshawar on 25 April 1992, under which the transitional arrangements were worked out.[36]

OBSTACLES TO STABILITY

The changes in the international environment over the past two years have removed the most stubborn obstacle in the way of negotiating political settlements of regional conflicts in places like Afghanistan and Angola, but the question remains whether deep-rooted indigenous polarizations and their inherent contradictions will disappear with the end of the cold war.

In Afghanistan, the long-term conflict has changed the traditional social and political order of the country. The most important development is internal political fragmentation. The Mujahideen parties are riven with fierce political and personal differences. Their internal divisions and the absence of democratic political institutions will doubtless complicate the process of maintaining a stable governing coalition. This instability in turn will contribute to ongoing instability in relations with and among other states of the region.

NOTES

1 *See,* Francis Fukuyama, *The Security of Pakistan: A Trip Report* (Santa Monica, California: Rand Corporation, September 1980).

2 Rais Ahmad Khan, 'US Policy Towards Afghanistan', *Pakistan Horizon,* Vol. 40, No. 1, 1987, pp. 65-79. *See also,* Milton Leitenberg, 'United States foreign policy and the Soviet invasion of Afghanistan', *Arms Control,* Vol. 7, No. 3, December 1986, pp. 271-94.

3 Hafiz Malik, 'Soviet Intervention in Afghanistan and its Impact on Pakistan's Foreign Policy,' in Malik, (ed.), *Soviet-American Relations with Pakistan, Iran and Afghanistan* (New York: St. Martin's Press, 1987), p. 131.

4 Stephen P Cohen, 'South Asia after Afghanistan', *Problems of Communism,* January-February 1985, pp. 18-31.

5 Tom Rogers, 'Afghan Refugees and the Stability of Pakistan', *Survival,* Vol. 29, No. 5, September-October, 1987, pp. 416-29.

6 Barbara Crossette, 'US Helps Russians Search for Missing in Afghanistan,' *New York Times,* 23 June 1992, p. 3.

7 India has refused to sign the NPT or to open up some of its facilities to international safeguards. She has continued production of weapons-grade nuclear materials and testing of ballistic missiles. In these moves Pakistan sees a threat to her security and has reacted by acquiring nuclear technology with the potential to produce weapons-grade materials.

8 In October 1990, the US President withheld certification to Congress that Pakistan does not possess nuclear weapons. The disbursement of $573 million earmarked for the fiscal year 1990-1 was suspended. *Muslim,* 5 October 1990.

9 'Pakistan Looks North,' *The Economist* (London), 22 February 1992, p. 29.

10 These views were expressed to Sardar Assef Ali, then Pakistan's minister of state for economic affairs, when he met the leaders of the Central Asian states in early 1991. *See, Muslim,* 7 January 1992.

11 *See,* statement of Benazir Bhutto, *News,* 12 April 1991.

12 *See,* the comments of Pakistani analysts F U Babar, 'Afghan Problem Nearing Solution?' *Muslim,* 14 June 1991; Maleeha Lodhi, 'A Rare Window for Peace in Afghanistan' *News,* 2 April 1991; and Nasim Zehra, 'Afghanistan; Unchanging Realities and New Realizations?', *News,* 19 May 1991.

13 After weeks of protests organized by the opposition Democratic Party and Islamic Renewal Party, President Rakhman N Nabiyev went into hiding on 6 May when the rebels took over the Parliament and radio buildings in Dushanbe. *New York Times,* 8 May 1992, p. 11. Later they agreed to a coalition government, retaining Nabiyev as President until elections in December. The opposition took control of the key ministries of defence, interior, foreign affairs, agriculture, trade and economy. *New York Times,* 12 May 1992, p. 5.

14 Acting Deputy Foreign Minister Hamid Karzai's statement, Hong Kong, AFP, 20 May 1992, in FBIS-NES-92-099, 21 May 1992, p. 35.

15 At the end of the summit, they announced construction of railways, roads and telecommunications and co-operation in the development of mineral and energy resources. See editorials in *Dawn,* 12 May 1992; *Nation,* 12 May 1992.

16 *Time* (Magazine), 27 April 1992, pp. 45-6.

17 Elaine Sciolino, 'Gandhi Faults Islamic Rule for Kabul', *New York Times,* 11 June 1988.

18 See an official statement of the Indian representative made at the emergency session of the United Nations General Assembly on 11 January 1980. UN General Assembly, Sixth Emergency Special Session, *Provisional Verbatim Record of the Third Meeting,* document A/ES-6PV, 3, General Assembly, United Nations, 11 January 1980.

19 *Foreign Affairs Record* (New Delhi), January 1980, p.19.

20 Yossef Bodansky, 'New Pressures on Key Indian Borderlands,' *Janes' Defence Weekly* (London), 30 April 1988, p. 840

21 For a fuller account of Indian policy, *see,* Thomas Perry Thornton, 'India and Afghanistan', in Theodore L Eliot, Jr., and Robert L Pfaltzgraff, Jr., (eds.), *The Red Army on Pakistan's Border: Policy Implications for the United States* op. cit., pp. 44-70.

22 *See,* a report by Salamat Ali, 'India Gets Involved: Gandhi Signals Support for Najibullah's Government', *Far Eastern Economic Review,* 19 May 1988, p. 35. Indeed, India was the first country outside the Soviet bloc to invite Najibullah to visit after the Soviet withdrawal.

23 *See,* Ahmad Shah Masud's interview, *Dawn,* 24 April 1992.

24 Radio Delhi, 31 December 1988, in FBIS-NES, 3 January 1989, p. 43.

25 On the origins of the Kashmir dispute *see,* Alastair Lamb, *Kashmir: A Disputed Legacy* (Hertingfordbury: Rexford Books, 1991); and Prem Nath Bazaz, *History of the Struggle for Freedom in Kashmir* (New Delhi: Kashmir Publishing Co., 1954).

26 The Agreement put an end to the long period of conflict. Both countries pledged an era of 'peaceful coexistence' and not to unilaterally change the cease-fire line in Jammu-Kashmir. *See,* Lawrence Ziring, 'Bhutto's Foreign Policy,' in Henry Korson, (ed.) *Contemporary Problems of Pakistan* (Leiden: E J Brill, 1974), pp. 56-80.

27 Pakistan has been accusing India of violating human rights in Kashmir, while India has alleged that Pakistan is supporting the freedom fighters. Pakistan raised the Kashmir issue at the NAM foreign ministers' meeting held in May 1992 in

Indonesia, and has threatened to take up the issue before the UN Security Council. *Muslim,* 16 May 1992.

28 For example, the former Indian Prime Minister V P Singh told parliament that India should be 'psychologically prepared for war,' *Financial Times* (London), 11 April 1990.

29 *Asiaweek* (Hong Kong), 9 February 1990, p. 28; and *Financial Times,* 13 February 1990.

30 Iranian foreign minister's statement, Radio Tehran, 4 April 1990, in FBIS-NES, 5 April 1990, p. 40.

31 Radio Tehran, 13 March 1990, in FBIS-NES, 15 March 1990, p. 58.

32 Professor Burhanuddin Rabbani's interview in *Ukaz* (Jiddah), 31 March 1990, translated in FBIS-NES, 5 April 1990, p. 36.

33 Author's interviews with Afghan resistance leaders in Peshawar, January 1990. Tehran was reportedly more uncompromising on this issue than were some of the Shiite resistance groups themselves.

34 *Muslim,* 26 November 1990.

35 Mushahid Hussain, 'An 'Afghan Solution' for Peace', *Nation,* 27 April 1992.

36 Donatella Lorch, 'Rebels Agree on Interim Rule for Kabul', *The New York Times,* 25 April 1992.

Chapter 10

Conclusion

Restructuring peace and normalcy in a society that has experienced the devastation and horror of a Soviet interventionary war lasting over nine years, followed by a seemingly endless period of civil strife, may not be easy, or indeed possible. What makes this task even more difficult is the fragmentation of the traditional power structure and the emergence of localized centres of authority, often antagonistic to each other. No credible social or political institution or nationally acceptable leadership is available within the country to initiate and promote political reconciliation.

Having lost its traditional balance, and suffering from a tortured social psyche, Afghanistan presents a confused political and military situation in which each group appears to be attempting a unilateral solution to the national crisis while claiming the support of the majority. Ironically, the great majority of the Afghans, who have borne the most devastating effects of the war, are not only left out of the political process, but are also held hostage to the confrontation between competing groups, the striking feature of whose political and military strategies is the elimination, or at best exclusion, of the other groups. Dangerously, none of the adversaries has given up the idea of war as the instrument of political settlement.

Rivalry among the Afghan political groups, which has intensified since the collapse of the Marxist regime, is as old as the resistance itself. Political fragmentation and the total absence of a unified national leadership impeded the development of the Afghan resistance into a truly national liberation movement. In fiercely protecting the independent status of their groups, the Afghan resistance leaders completely ignored the vital and strategic linkages between a centralized political organization and co-ordinated military operations. For an outside observer it was intriguing to note the deep factional rivalries among the Mujahideen groups in the face of a vastly powerful enemy in the military sense, while having the common goal of defeating the Marxist regime. For several reasons the politics of the resistance were characterised by bitter discord rather than co-operation. In understanding the strife within the resistance and its effects on the contemporary politics of Afghanistan, one may refer to factors such as tribal affiliations, social stratification, particularism of the Afghan individual, and the quest for group identity. Underdeveloped organizational structures and the absence of charismatic national personalities also help to explain the fractious nature of the resistance. Furthermore, social forces and the circumstances of the evolution of the Afghan resistance have perpetuated differences which may roughly be characterized as doctrinal, political, and personal.

A widely-held notion that all Mujahideen groups are essentially Islamic, and therefore fundamentalist, may lead to erroneous conclusions. Those familiar with the politics of Islam in Muslim societies would know how various political parties with religious orientations are doctrinally opposed to each other. It would make common sense to raise the banner of Islam for its political value both in mobilizing support as well as in legitimising leadership in the cultural context of the Afghan society, but this should not obscure the fact that there are fundamental differences on the interpretation of Islam and the definition of its central principles. Divisions on the application of Islam to the state and politics are more manifest than those regarding ritual and individual practices.

The distinctions between the Islamists and the traditionalists not only project different visions of the future Afghan state but also generate irreconcilable communal hostility. This sentiment has long divided Afghan society. It has to some extent been the decentralized, autonomous and independent growth of the religious establishment which has produced separate organizations and leaders. The people, individually or as a group, have shown loyalty to their immediate tribal, ethnic or religious organization. Any coalescing of the groups participating in various networks would depend to a great extent on whether or not the leaders co-operated closely. Largely in order to seek political recognition and access to patronage, the leaders of various religious groups have maintained their independence. Such a posture was also not without economic dividends, as during the war of resistance weapons and foreign economic aid were allocated to these groups for distribution among their guerrillas and client population. That is why even after the ouster of the Marxist regime differences have remained as strong as during the war.

The second category of differences relates to political issues. Social class origin and Islamist-traditionalist orientations have shaped divergent views on politics, state and society. The conflicting political attitudes within the Mujahideen parties were more or less carried over from their social class basis and the political roles they had held before the war. The Islamist leaders and workers, having been on the fringes of political power, represent the social ethos of a rising educated middle class, cutting across rural, tribal and ethnic affiliations. The major focus of their political programme, though less salient during the war, has been on ending the political domination of the Afghan aristocracy, along with Islamizing society. On the other hand the traditionalist, nationalist leaders were part of the privileged class, and had a natural interest in aligning with the forces that would preserve the social and political status quo. The destabilization of the old social order by the war and the vastly different Afghan reality of today may have influenced their political thought, but the traditionalists are suspected by the Islamists of harbouring the desire to restore the old social relations and political arrangements.

The third category of differences might be explained in terms of personality clashes and rivalry for personal recognition and leadership. Much of the infighting among the Mujahideen groups is a reflection of personal antagonisms, which is generally extended to the level of ideology. Mutual suspicions grew more widely with the autonomous control of territories by the various groups inside Afghanistan. The competition for political dominance prompted the internal commanders who were loosely aligned to the Peshawar-based political parties to outflank each other. Sabotage, assaults and ambushes of a serious nature occurred among the rival groups. The squabbles among the resistance parties, which had been to some extent sidelined by the presence of the Soviet forces, and were muted somewhat by the survival of the Marxist regime, have become increasingly public and violent. The first round of battles occurred over who would control Kabul and fill the power vacuum left by the sudden fall of Najibullah in April 1992. Although the Peshawar Accord, signed by a majority of the Mujahideen parties, paved the way towards a peaceful transition, the presence of the Uzbek militia in Kabul, supported by the guerrillas of Ahmad Shah Masud, led to bloody clashes with the forces of Hikmatyar. This resulted in hundreds of deaths and the expulsion of the Hizb from its positions in the city. The effects of these events have been disastrous for the unity of the transitional government. The two most organized parties, with identical views on Islamism, became locked in a bitter conflict. The leaders traded accusations of collaborating with the functionaries of the defunct Marxist regime, while their guerrillas fought battles against each other. The agreement between Masud and Hikmatyar arranged by their common foreign friends in May, the transfer of power to Rabbani and the Hizb's decision to join his government have eased these tensions. Whether or not the accord and a new partnership will pave the way towards stable relations and a genuine reconciliation between the two rivals is yet to be seen; much will depend on how the political forces realign.

While assessing the military capabilities and political programmes of the contending Mujahideen groups, the role of civil society and its institutions should not be ignored. The moot question is: how likely is the Afghan population to support the efforts of the Mujahideen parties if they decide to plunge the country into yet another round of war? Having suffered the enormous devastation and dislocation caused by the Soviet war, ordinary Afghans are no longer enthusiastic about military solutions; rather, they are tired and desperate to end the uncertainty and restore peace. But ordinary Afghans find themselves torn between the competing groups. There is no independent institution with enough credibility in civil society to be able to disengage fiercely antagonistic adversaries, particularly of the ethnic variety. The outcome of the conflict among the Mujahideen parties will depend partly on which side succeeds in forging organized links with society, convincing it about the truth of its ideals. The more prolonged this process, the more tragic the situation for the war-ravaged country.

A failure to build a political system based on consensus, participation, and power-sharing would force the central state to rely on the use of security forces. The Afghan experience over the last fifteen years amply demonstrates the fact that recourse to the use of violence by the state is no substitute for political accommodation with the opponents. Needless to say, any policy of subjugating or eliminating political foes with the help of the army would undermine the power of the political leaders, who have yet to consolidate their authority. During the entire period of civil strife, the Afghan armed forces have never been immune from the influence of political forces. This is evidenced by the fact that the Marxists rode to power on the backs of military officers who were secretly working for their party. Likewise, some of the Mujahideen groups maintained underground networks within both the armed forces and the intelligence agencies, using tribal and ethnic links. Since the fall of the Marxist regime, a sizeable number of army officers and soldiers have defected to Mujahideen parties along ethnic and sectarian lines. Reorganizing or raising a truly national army would be problematic, and such efforts would be further hampered by the political rift between the Mujahideen parties. In a society that is divided by narrow loyalties, *coup* attempts in the future cannot be ruled out completely, because the Afghan armed forces have become politicized and, in the final analysis, are but a reflection of the fragmentation of Afghan society.

The political fragmentation of Afghanistan along ethnic lines presents a serious challenge to the territorial unity of the country. The important thing to recognize is that the conflict has brought to the fore new social and political forces. The most important change is the strengthening of the ethnic identity. Non-Pushtun minority groups such as the Hazaras, Tajiks, Turkmens and Uzbeks have acquired a greater sense of empowerment by establishing autonomous administrative networks. The Soviets paid greater attention to the development of the northern areas, which are essentially non-Pushtun. The military power of the Uzbek militia has strengthened speculations that a *de facto* parallel government might be formed in these areas. This is a new political reality which the Pushtun majority has to recognize and accommodate in order to maintain a unified Afghanistan. With the emergence of Mazar-e-Sharif as an important centre of political and military power, northern Afghanistan and its Tajik-Uzbek coalition cannot be ignored in any future power arrangement. Their exclusion from power, or the apportionment of a lesser role than demanded by their numbers and strategic position, would perpetuate conflict.

Ethnic war in Afghanistan would also complicate the role of the regional states. As long as the conflict remained primarily between the Marxist regime and the resistance groups, Tehran and Islamabad were accepted by the majority of the Afghans as benign and true supporters of their struggle, but with the departure of the Marxist regime from the scene, and ethnic polarization, the whole structure of conflict is in the process of change. We might discern a trend towards a new pattern of alliances along ethnic and regional lines that would shape the struggle for power. This would make the role of the regional states more complex and certainly less manageable. The divisions and infighting, and obviously the quest for dominance among the Afghan parties, would make it practically impossible for Iran and Pakistan to maintain a neutral stance. As they have played favourites in the past, it would be hard for them to escape stigmatization from those groups which might find themselves weaker or left out of the Afghan power struggle.

There is no doubt that Tehran and Islamabad have a strategic interest in the political stability of Afghanistan. Both are looking beyond Afghanistan to the markets of the Central Asian states. For Pakistan, only a stable and peaceful Afghanistan would guarantee uninterrupted access. But this also points to the danger of becoming enmeshed in the internal affairs of Afghanistan. Unfortunately, all the neighbouring countries have competing interests in Afghanistan due to the overlapping of both ethnic groups and political interests.

There is a greater fear that balkanization of the country might tempt the neighbouring states to carve out their spheres of influence by supporting favourite groups. Geography, history, and ethnicity might guide the choice of territories and groups compatible with Iran's and Pakistan's interests. In a territorial divide along ethnic lines, the northern minorities, who share geography, culture and language with the peoples of the contiguous Central Asian states, might be attracted by the idea of seeking alignments with them against the resurgence of Pushtun dominance. Similarly, the fear of a hegemonic alliance of the Sunni groups would drive the Shiite population of the Hazarajat and the adjoining region of Hirat into the Iranian fold. Tehran's interest in the autonomy of the Shiite areas is well-established. The influence of Persian culture and language is another factor that would strengthen the geopolitical interest of Tehran. The Pushtuns living in the eastern and southern parts of Afghanistan have tribal, ethnic, and family ties with the Pushtuns of the Balochistan and Frontier provinces of Pakistan. However, it would be impossible for Pakistan to establish or maintain any degree of stable influence over the Pushtuns, for two reasons. First, the Pushtuns are more divided along tribal and political lines than any other ethnic group. Second, the Pushtun identification with Afghanistan and militant nationalism would definitely complicate any attempt by Pakistan to dominate them. Conversely, their infighting and common desire to unify the country might continue to drag Pakistan into their struggles. At this point it is difficult to predict the outcome of ethnic and political conflict among the Afghan groups supported by outside powers. But one of the serious and most likely implications of the involvement of regional powers would be a sharpening of tensions inside Afghanistan.

It is evident now that none of the Mujahideen parties can alone establish legitimacy or normalize the situation in Afghanistan. Is there a way to end their conflict? How can normalcy and peace be restored in the country? What should be the essential ingredients of a national reconciliation strategy? The vital question today is how to establish a legitimate central authority on the basis of a wide political consensus that would create the conditions necessary for the restoration of normal life. Such a goal might appear illusive in view of the prevailing climate of hostilities and mutual antagonism among the various Afghan factions, but no political settlement in a real sense will hold unless it is directed towards reconstituting a broad and legitimate government.

What we propose here are three essential steps. First, the formation of a broad-based national government. Significant progress has been made with the transfer of power to Rabbani as proposed in the Peshawar Accord. He is more likely to draw wide support, but he will succeed in keeping the coalition together only if he moves with a consensus of the leadership council, which comprises the heads of all the Mujahideen parties. While the public postures of the political parties remain poles apart on some of the fundamental issues, e.g. autonomy for ethnic and regional groups, one need not rule out an eventual compromise.

Second, the drafting of a constitution by a commission of jurists and legal experts. Representation may be given to all parties on this body so that differences on the form of the government, the ideology of the constitution and, more importantly, the redrawing of the powers of the central government and of the regions or provinces, are settled before its ratification. No progress towards the further political development of the country can be expected until and unless this basic document is prepared and immediately put into operation. Otherwise, the political re-ordering would lack legal justification and political support, leaving the door wide open to confrontation. A constitution may not guarantee peace or orderly politics in the fragmented society of Afghanistan, but that is where the quest for political order must begin.

Third, the conducting of free and fair elections. There are serious obstacles and logistical problems in the way of general elections in the present state of political anarchy in Afghanistan, but, in our judgement, there is no long-term alternative to this institution for a peaceful political change and for giving legitimacy to the rulers. Afghanistan has to start somewhere on this road, no matter how bumpy and rough it is. When there are so many contenders for political power, electoral procedures are the only means to determine the legitimacy of their claims. The process will need to be kept free of all the external actors who have been directly or indirectly involved in the Afghan conflict. One reason why the resolution of the conflict is so complex is the clashing interests of very diverse elements. Only understanding, reconciliation, and a spirit of compromise among them would produce an environment conducive to peace in Afghanistan.

The devising of peace strategies for a fragmented society can be frustrating due to the multipolarity of confrontations. What further complicates the problem is the creation and exploitation of a geopolitical situation by a large number of actors. In view of the complexity of interests behind the support given to the various parties in the Afghan conflict, an integrated and concerted approach to restructuring political legitimacy is necessary. The search for a settlement must be guided by the free choice of the Afghans in organizing their own political future. While those foreign powers genuinely interested in a political settlement can help by creating the conditions for national reconciliation, success will depend on how the Afghans play out the game of political power. The destruction of the national assets of the country and the widespread human suffering and dislocation call for collective action. The legacy of the Soviet war plus civil strife, which has made the path of transition uncertain, will persist if the Afghan leaders fail to make a new social contract, a contract for pluralism, political co-existence, sharing, and accommodation. What the Afghans and their supporters need to realize is that, in a rapidly changing international environment, the settlement of domestic political disputes through war is becoming counter-productive and obsolete. The sooner this realization occurs, the better will be the chances for political stability in Afghanistan.

Bibliography

Adamec, Ludwig W, *Afghanistan's Foreign Affairs to the Mid-Twentieth Century. Relations With the USSR, Germany, and Britain.* Tucson: University of Arizona Press, 1974.

______, *Afghanistan, 1900-1923: A Diplomatic History.* Berkeley: University of California Press, 1967.

Afghan Demographic Studies, *A Provisional Gazetteer of Afghanistan*, Vol. 2. Kabul: Central Statistics Office, Prime Ministry. Report Series No. 1, 1975.

Afghanistan National Fatherland Front, *Documents of the Founding Congress.* Kabul, 15-16 June, 1981. New Delhi: Communist Party of India Publication, 1981.

Afghanistan, Ministry of Planning, *Statistical Pocketbook of Afghanistan*, August 1972.

Afghanistan: *The Target of Imperialism*, Kabul: Party Printing Press, 1983.

Ahady, Anwar-ul-Haq, 'Conflict in Post-Soviet Occupation Afghanistan'. *Journal of Contemporary Asia*, Vol. 21, No. 4 (1991).

Ahmad, Eqbal, 'Revolutionary Warfare and Counter-Insurgency', in Norman Miller and Roderick Aya, (eds.) *National Liberation: Revolution in the Third World.* New York: Free Press, 1971.

Ahmad, Feroz, 'Afghan Developments X-rayed'. *The Marxist Review.* Calcutta, Vol. 24, No. 3 (September 1980).

Ahmed, Akbar S, *Millennium and Charisma among Pathans.* London; Routledge and Kegan Paul, 1976.

______, *Pakhtun Economy and Society.* London: Routledge and Kegan Paul, 1980.

______, 'The Impact of the Afghan Refugees on Ethnicity and Politics in Baluchistan (Pakistan)'. *Central Asian Survey*, Vol. 9, No. 3 (1990).

______, 'Honour and Power in Afghanistan: Political Significance of Cultural Models in Society'. *Strategic Studies*, Vol. 4, No. 4 (Summer 1981).

Akbar, Mansoor, 'Revolution and Counter-Revolution in Afghanistan'. *Journal of Contemporary Asia*, Vol. 18, No. 4 (1988).

Akhramovich, Roman T, *Outline History of Afghanistan After the Second World War.* Moscow: Nauka Publishing House, 1966.

Alavi, Hamza, 'Peasant Classes and Primordial Loyalties'. *Journal of Peasant Studies*, Vol. 1, No. 1 (1973).

Alexiev, Alex, *Inside the Soviet Army in Afghanistan:* Santa Monica: Rand Corporation, 1988.

______, 'US policy and the War in Afghanistan'. *Global Affairs*, Vol. 3 No. 1 (Winter 1988).

Ali, Mehrunisa, *Pak-Afghan Discord: A Historical Perspective, Documents 1853-1979.* Karachi: Pakistan Study Centre, Karachi University, 1990.

Ali, Sardar Shaukat, *Inqelab-e-Afghanistan.* (Revolution in Afghanistan) Lahore: Peoples Publishing House, 1987.

Allan, Peirre, and Albert A Stahel, 'Tribal Guerrilla Warfare Against a Colonial Power'. *Journal of Conflict Resolution,* Vol. 27, No. 4 (1983).

Allworth, Edward, (ed.) *Central Asia: Bukhara and Khiva, 1865-1924.* Cambridge: Harvard University Press, 1968.

Amin, A Rasul, 'A General Reflection of the Stealth Sovietisation of Afghanistan'. *Central Asian Survey,* Vol. 3, No. 1 (1984).

———, 'The Saur Revolution,' *WUFA, Quarterly Journal of Writers Union of Free Afghanistan,* Vol. 1, No. 2, (Jan-March 1986).

Amin, Hafizullah, 'Afghanistan's Revolution Today'. The last interview Mr Amin gave before he was ousted. *Vikrant,* Vol. 4 (January 1980).

Amin, Tahir, 'Afghan Resistance: Past, Present and Future'. *Asian Survey,* Vol. 24 (April 1984).

———, *Afghanistan Crisis: Implications and Options for Muslim World, Iran and Pakistan.* Islamabad: Institute of Policy Studies, 1982.

Amstuts, J Bruce, *Afghanistan: The First Five Years of Occupation.* Washington, D.C.: National Defence University Press, 1986.

Anderson, John W, and Richard F Strand, (eds.) *Ethnic Processes and Intergroup Relations in Contemporary Afghanistan.* Occasional Paper No. 15. New York: Asia Society, Afghanistan Council, 1978.

———, 'There are No Khans Anymore: Economic Development and Social Change in Tribal Afghanistan'. *Middle East Journal,* Vol. 32, No. 2 (1978).

———, 'Tribe and Community among Ghilzai Pashtun'. *Anthropos,* Vol. 70 (1975).

Ansari, Javed, 'Islam Finds Marxism Wanting'. *Arabia: The Islamic World View,* Vol. 1, No. 1 (1981).

Anwar, Raja, *Tragedy of Afghanistan: A Firsthand Account.* (translated from the Urdu by Khalid Hasan). London: Verso, 1988.

Arnold, Anthony, *Afghanistan's Two-Party Communism: Parcham and Khalq.* Stanford: Hoover Institution Press, Stanford University, 1983.

———, *Afghanistan: The Soviet Invasion in Perspective.* Stanford: Hoover Institution Press, Stanford University, 1982.

Aya, Roderick, 'Theories of Revolution Reconsidered'. *Theory and Society,* Vol. 8, No. 1 (1979).

Azmi, Muhammad R, 'Soviet Politico-Military Penetration in Afghanistan 1955 to 1979'. *Armed Forces and Society,* Vol.12, No. 3 (Spring 1986).

Azoy, G Whitney, *Buzkashi: Game and Power in Afghanistan.* Philadelphia: University of Pennsylvania Press, 1982.

'B M', 'The Present Situation in the Hazarajat'. *Central Asian Survey,* Vol. 1, No. 1 (1982).

Bain, R B, *Afghanistan Will not Die.* Calcutta: Satyam Shivam, 1982.

Baitenmann, Hela, 'NGOs and the Afghan War: the Politicisation of Humanitarian Aid'. *Third World Quarterly,* Vol. 12, No. 1 (January 1990).

Barfield, Thomas J, *The Central Asian Arabs of Afghanistan.* Austin: University of Texas Press, 1981.

Barth, Fredrik, *Political Leadership among Swat Pathans.* London: Athlone Press, 1959.

Basic Lines of the Revolutionary Duties of the Government of the Democratic Republic of Afghanistan, Kabul: Government Printing Press, 9 May 1978.

Beattie, Hugh, 'Kinship and Ethnicity in the Nahrin Area of Northern Afghanistan' *Afghan Studies*, Nos. 3/4 (1982).

Bhargava, G S, *South Asian Security after Afghanistan.* Lexington, Mass: D C Heath and Co., 1983.

Bhatty, I Z and L Berouti, 'A Development Strategy for Afghanistan: Lessons of an Employment Policy Mission'. *Pakistan Development Review*, Vol. 19, No. 4 (1980).

Biographical Accounts of Chiefs, Sardars and Others of Afghanistan, Calcutta: Foreign Office, India, 1888.

Biography of Nur Mohammad Taraki, Kabul: Afghan Government Publication, August 1978.

Blank, Stephen, 'Soviet Low-Intensity Conflicts: An Agenda for Research'. *Conflict*, Vol. 9, No. 1 (1989).

Bociurkiw, Bohdan R, 'Changing Soviet Image of Islam: Domestic Scene'. *Journal of the Institute of Muslim Minority Affairs*, Vol. 2 No. 2, Vol. 3 No. 1 (1980-81).

Bodansky, Yossef, 'New Pressures on Key Indian Borderlands'. *Janes' Defence Weekly*, 30 April 1988.

Bokhari, Imtiaz H, 'Evolution of Dual Negotiation Process: Afghanistan'. *Annals of the American Academy of Political and Social Sciences*, Vol. 518 (November 1991).

Bradsher, Henry S, *Afghanistan and the Soviet Union.* Durham: Duke University Press, 1985.

_______, 'Stagnation and Change in Afghanistan'. *Journal of South Asian and Middle Eastern Studies*, Vol. 10, No. 1 (Fall 1986).

Brezhnev, Leonid, 'On the Afghan Situation'. *Reprints from the Soviet Press*, Vol. 30, No. 2.

Burki, Shahid Javed, (ed.) *Pakistan Under the Military: Eleven years of Ziaul Haq.* Boulder: Westview Press, 1991.

Canfield, Robert L, 'Soviet Gambit in Central Asia'. *Journal of South Asian and Middle Eastern Studies*, Vol. 5, No. 1 (1981).

_______, 'Afghanistan: The Trajectory of Internal Alignments'. *Middle East Journal*, Vol. 43, No. 4 (Fall 1989).

Caroe, Sir Olaf, 'Afghanistan: The After Effects of Partition'. *The Round Table*, No. 278 (April 1980).

_______, *Soviet Empire: The Turks of Central Asia and Stalinism.* 2nd ed. New York: St. Martin's Press, 1967.

_______, *The Pathans; 550 BC-AD 1957.* Macmillan, 1965.

Chaffetz, David, 'Afghanistan in Turmoil'. *International Affairs*, Vol. 56, No. 1 (1980).

———, 'Afghanistan, Russia's Vietnam'. New York: Asia Society, Afghanistan Council. Special Paper No. 4, 1979.

Chaliand, Gerard, *Report from Afghanistan*. Translated by Tamar Jacoby. New York: Penguin Books, 1982.

Charpentier, C J, 'One Year After the Saur Revolution'. *Afghanistan Journal*, Vol. 6, No. 4 (1979).

Cheema, Pervaiz Iqbal, 'Impact of the Afghan War on Pakistan'. *Pakistan Horizon*, Vol. 41, No. 1 (January 1988).

Chu, Solomon, et al. *National Demographic and Family Guidance Survey of the Settled Population of Afghanistan*, Vol. 1: Demography. Kabul: Government of Afghanistan and USAID, 1975.

Cohen, Stephen P, 'South Asia after Afghanistan'. *Problems of Communism*, (January-February 1985).

Collins, Joseph J, 'Afghanistan: The Empire Strikes Out'. *Parameters, Journal of the U.S. Army War College*, Vol. 12, No. 1 (1982).

———, 'Soviet Policy toward Afghanistan'. *Proceedings of the Academy of Political Science*, Vol. 36, No. 4 (1987).

Cronin, Richard P, 'Afghanistan in 1988: Year of Decision'. *Asian Survey*, Vol. 29, (February 1989).

———, *Afghanistan: Soviet Invasion and US Response*. Washington, D.C.: Congressional Research Service, Library of Congress, updated 24 November, 1981.

Cynkin, Thomas M, 'Aftermath of the Saur Coup'. *Fletcher Forum*, Vol. 6, No. 2 (Summer 1992).

Daley, Tad, 'Afghanistan and Gorbachev's Global Foreign Policy'. *Asian Survey*, Vol. 29 (May 1989).

Dar, Saeeduddin Ahmad, (ed.) *Selected Documents on Pakistan's Relations with Afghanistan 1947-85*. Islamabad: National Institute of Pakistan Studies, Quaid-i-Azam University, 1986.

Dastarac, Alexandre, and M Levant, 'What Went Wrong in Afghanistan?' *MERIP Reports*, Vol. 10, No. 6, (July-August 1980).

Davies, C Collin, *The Problem of the North-West Frontier 1890-1908*. Cambridge: Cambridge University Press, 1932.

Davison, Kenneth L Jr., 'The Geopolitics of the Soviet Withdrawal from Afghanistan'. *Strategic Review*, Vol. 18, No. 1 (Winter 1990).

Dekmejian, R Hrair, 'The Anatomy of Islamic Revival: Legitimacy Crisis, Ethnic Conflict, and the Search for Islamic Alternatives'. *Middle East Journal*, Vol. 34, No. 1 (1980).

Djabarov, A, 'Afghanistan: The Revolutionary Process is Irreversible'. *Soviet Military Review*, (March 1981).

Dorronsoro, Gilles and Chantal Lobato, 'The Militia in Afghanistan'. *Central Asian Survey*, Vol. 8, No. 4 (1989).

Dupree, Louis, et al. *Afghanistan in the 1970s*. New York: Praeger Publishers, 1974.

———, 'A Note on Afghanistan 1974'. *AUFS Reports*, South Asia Series, No. 6, 1974.

_______ , *Afghanistan.* Princeton: Princeton University Press, 1980.

_______ , 'Afghanistan: 1980, The World Turned Upside Down'. *AUFS Reports,* Asia, No. 37, 1980.

_______ , 'Islam in Politics: Afghanistan' *Muslim World,* Vol. 16, No. 4 (1966).

_______ , 'Militant Islam and Traditional Warfare in Islamic South Asia'. *AUFS Reports,* Asia, No. 21, 1980.

_______ , 'Red Flag over the Hindu Kush, Part I: Leftist Movements in Afghanistan'. *AUFS Reports,* Asia, No. 44, 1979.

_______ , 'Red Flag over the Hindu Kush, Part II: The Accidental Coup, or Taraki in Blunderland'. *AUFS Reports,* Asia, No. 45, 1979.

_______ , 'Red Flag Over the Hindu Kush, Part III: Rhetoric and Reforms, or Promises! Promises'. *AUFS Reports,* Asia, No. 23, 1980.

_______ , 'Red Flag over the Hindu Kush, Part IV: Foreign Policy and Economy'. *AUFS Reports,* Asia, No. 27, 1980.

_______ , 'Red Flag over the Hindu Kush, Part V: Repression, or Security Through Terror, Purges I-IV'. *AUFS Reports,* Asia, No. 28, 1980.

_______ , 'Red Flag over the Hindu Kush, Part VI: Repressions, or Security Through Terror, Purges IV-VI'. *AUFS Reports,* Asia, No. 29, 1980.

_______ , 'The Democratic Republic of Afghanistan'. *AUFS Reports,* Asia, No. 32, 1979.

_______ ,'The First Anglo-Afghan War and the British Retreat of 1842: The Functions of History and Folklore'. *East and West,* Vol. 26, Nos. 3-4 (1977).

_______ ,'Toward Representative Government in Afghanistan, Part I: The First Five Steps'. *AUFS Reports,* Asia , No. 1, 1978.

_______ ,'Toward Representative Government in Afghanistan, Part II: Steps Six Through Nine—and Beyond?' *AUFS Reports,* Asia , No. 14, 1978.

Dupree, Nancy Hatch, *Revolutionary Rhetoric and Afghan Women.* New York: Asia Society, Afghanistan Council, Occasional Paper No. 23, 1981.

Eberhard, Wolfram, 'Afghanistan's Young Elites'. *Asian Survey,* (February 1962).

Eliot, Theodore L Jr., and Robert L Pfaltzgraff Jr., (eds.) *Red Army on Pakistan's Borders.* Washington, D.C.: Pergamon-Brassey's, 1986.

_______ , 'Afghanistan in 1990: Groping Toward Peace?' *Asian Survey,* Vol. 31, (February 1991).

Elphinstone, M, *An Account of the Kingdom of Caubul: and Its Dependencies in Persia, Tartary and India,* 2 Vols. 3rd edition (Karachi Oxford University Press, 1972).

Embree, Ainslee T, (ed.) *Pakistan's Western Borderlands.* Durham, N.C.: Carolina Academic Press, 1977.

Engert, C van H, *A Report on Afghanistan.* Washington, D.C.: Government Printing Office, 1924.

Farr, Grand M and John G Merriam, (eds.) *The Afghan Resistance; The Politics of Survival.* Boulder, Colo.: Westview Press, 1987.

Ferdinand, K, and M Mozaffari, (eds.) *Islam: State and Society.* London: Curzon, 1988.

_______ , 'Nomad Expansion and Commerce in Central Afghanistan'. *Folk,* No. 4 (1962).

Fischer, Michael J, 'Islam and the Revolt of the Petit Bourgeoisie'. *Daedalus* (Winter 1982).

Fletcher, Arnold, *Afghanistan: Highway of Conquest.* Ithaca: Cornell University Press, 1965.

Fromkin, David, 'The Great Game in Asia'. *Foreign Affairs* , Vol. 58, No. 4 (Spring 1980).

Fry, Maxwell J, *The Afghan Economy: Money, Finance, and the Critical Constraints to Economic Development.* Leiden: E J Brill, 1974.

Fukuyama, Francis, *The Security of Pakistan: A Trip Report.* Santa Monica: Rand Corporation, September 1980.

Gall, Sandy, *Behind Russian Lines; An Afghan Journal.* London: Sidgwick and Jackson, 1983.

Gankovsky, Yuri V, et al. *A History of Afghanistan.* Moscow: Progress Publishers, 1982.

———, 'International Relations, Peace and Security: Problem of South Asia in the Works of Soviet Orientalists'. *Strategic Studies,* Vol. 6, Nos. 2/3 (Winter/ Spring 1982-3).

Garg, J P, 'Russian Penetration in the Third World with Special Reference to Afghanistan'. *Indian Journal of Political Science,* Vol. 42, No. 4 (Oct-Dec. 1981).

Ghani, Abdul, *Review of the Political Situation in Afghanistan.* Lahore: Khosla Brothers, 1921.

Ghani, Ashraf, 'Islam and State-Building in a Tribal Society: Afghanistan 1880-1901'. *Modern Asian Studies,* Vol. 12, No. 2 (1978).

Gibbs, David, 'Does the USSR have a Grand Strategy? Reinterpreting the Invasion of Afghanistan'. *Journal of Peace Research,* Vol. 24, No. 4 (December 1987).

———, 'The Peasant as Counter-Revolutionary: The Rural Origins of the Afghan Insurgency'. *Studies in Comparative International Development,* Vol. 21, No. 1 (Spring 1986).

Goldman, Minton F, 'The Soviet Predicament in Afghanistan'. *Asian Thought and Society,* Vol. 37 (January 1988).

Government of Afghanistan, *Afghanistan 1990,* report prepared for United Nations Conference of the Least Developed Countries,1990.

Grachev, A S, (ed.) *The Undeclared War; Imperialism vs. Afghanistan.* Moscow: Progress Publishers, 1980.

Grassmuck, George, et al. *Afghanistan: Some New Approaches.* Ann Arbor: The University of Michigan, 1969.

Gregorian, Vatran, *The Emergence of Modern Afghanistan: Politics of Reform and Modernization, 1880-1946.* Stanford; Stanford University Press, 1974.

Griffith, William E, 'Superpower Relations after Afghanistan'. *Survival,* Vol. 22, No. 4 (July-August 1980).

Griffiths, John C, *Afghanistan: Key to a Continent.* Boulder, Colo.: Westview Press, 1981.

Gupta, Arvind, 'Soviet Military Intervention in Afghanistan in Perspective'. *Strategic Analysis,* Vol. 12, No. 2 (March 1990).

Halliday, Fred, 'Revolution in Afghanistan'. *New Left Review*. Vol. 112 (November-December 1978).

_______ , *The Making of the Second Cold War*. London: Verso, 1983.

_______ , 'War and Revolution in Afghanistan'. *New Left Review*, Vol. 119 (January-February 1980).

Hammond, Thomas T, *Red Flag over Afghanistan: The Communist Coup, the Soviet Invasion and the Consequences*. Boulder, Colo.: Westview Press, 1984.

Harrison, Selig S, 'A Breakthrough in Afghanistan'. *Foreign Policy*, No. 51, (Summer 1983).

_______ , 'After the Afghan Coup: Nightmare in Balochistan'. *Foreign Policy*, No. 32 (Fall 1978).

_______ , *In Afghanistan's Shadow: Baluch Nationalism and Soviet Temptations*. New York: Carnegie Endowment for International Peace, 1981.

_______ , 'Inside the Afghan Talks'. *Foreign Policy*, No. 72 (Fall 1988).

Havelock, Captain Henry, *Narrative of the War in Afghanistan in 1838-39*. London: Henry Colburn Publisher, 1840.

Heuer, Richards J, 'Analyzing the Soviet Invasion of Afghanistan: Hypotheses from Causal Attribution Theory'. *Studies in Comparative Communism*, Vol. 13, No. 4 (Winter 1980).

Hizb-i-Islami, *Aims of Hizb-i-Islami Afghanistan*. Peshawar: Hizb-i-Islami Afghanistan, d.n.

Huldt, Bo and Erland Jansson, (eds.) *The Tragedy of Afghanistan: The Social, Cultural and Political Impact of the Soviet Invasion*. London: Croom Helm, 1988.

Huntington, Samuel P, *Political Order in Changing Societies*. New Haven: Yale University Press, 1968.

Hyman, Anthony, *Afghanistan Under Soviet Domination: 1964-81*. New York: Saint Martin's Press, 1982.

_______ , 'The Afghan Politics of Exile'. *Third World Quarterly*, Vol. 9, No. 1 (January 1987).

Ilyinsky, Mikhail, *Afghanistan: Onward March of the Revolution*. New Delhi: Sterling Publishers, 1982.

Islam, Mohammad, 'US-Afghanistan Relations: A Historical Perspective'. *Pakistan Journal of American Studies*, Vol. 9, Nos. 1&2 (Spring/Fall 1991).

Jafri, H A S, 'Evolution of Afghanistan as an Independent State'. *India Quarterly*, Vol. 42, No. 2 (April-June 1986).

_______ , 'Soviet Intervention in Afghanistan'. *Indian Journal of Politics*, Vol. 14, Nos. 1-3 (1980).

Jettmar, Karl, 'Ethnological Research in Dardistan 1958'. *Proceedings of the American Philosophical Society*, Vol. 105, No. 1 (1961).

Jones, Schuyler, *Men of Influence in Nuristan: A Study of Social Control and Dispute Settlement in Waigal Valley, Afghanistan*. New York: Academic Press, 1974.

Kakar, M Hasan Kawun, 'The Fall of the Afghan Monarchy in 1973'. *International Journal of Middle Eastern Studies*, No. 9 (May 1978).

———, *Government and Society in Afghanistan: The Reign of Amir Abd al Rahman Khan.* Austin: University of Texas Press, 1979.

Kamrany, Nake M, 'Afghanistan Under Soviet Occupation'. *Current History*, (May 1982).

———, *Peaceful Competition in Afghanistan: American and Soviet Models for Economic Aid.* Washington, D.C.: Communication Service Corporation, 1969.

Karmal, Babrak, 'Afghan people will defend their revolution', an interview. *Soviet Review*, Vol. 18 (19 November 1981).

———, 'The Afghan People Will Stand Firmly on Guard of the Revolution'. *World Marxist Review*. Prague, Vol. 23, No. 4 (April 1980).

Karp, Aaron, 'Blowpipes and Stingers in Afghanistan, One Year Later'. *Armed Forces Journal International*, (September 1987).

Karp, Craig M, 'The War in Afghanistan'. *Foreign Affairs*, Vol. 64, No. 5 (Summer 1986).

Kasatkin, Dmitri, 'Protecting the Gains of the April Revolution'. *Asia and Africa Today*, No. 3 (1980).

Keddie, Nikki, (ed.) *Scholars, Saints, and Sufis; Muslim Religious Institutions Since 1500.* Berkeley: University of California Press, 1962.

Keiser, R Lincoln, 'Social Structure and Social Control in Two Afghan Mountain Societies'. Unpublished Ph. D. dissertation, University of Rochester, 1971.

Khalilzad, Zalmay, 'Afghanistan and the Crisis in American Foreign Policy'. *Survival*, Vol. 22, No. 4 (July-August 1980).

———, 'Moscow's Afghan War'. *Problems of Communism*, Vol. 35 (January-February 1986).

———, *The Return of the Great Game.* Santa Monica, CA.: California Seminar on International Security and Foreign Policy, September 1980.

———, 'The War in Afghanistan'. *International Journal*, Vol. 41, No. 2 (Spring 1986).

Khan, Rais Ahmad, 'US Policy Towards Afghanistan'. *Pakistan Horizon*, Vol. 40, No. 1 (1987).

Khan, Riaz M, *Untying the Afghan Knot: Negotiating Soviet Withdrawal.* Durham: Duke University Press, 1991.

Khan, Sultan Muhammad, 'Pakistani Geopolitics: The Diplomatic Perspective' *International Security*, Vol. 5, No. 1 (Summer 1980).

Khan, Sultan Muhammad Mir Munshi, *The Constitution and Laws of Afghanistan* London: Murray, 1900.

Klass, Rosanne, 'Afghanistan: the Accords'. *Foreign Affairs*, Vol. 66, No. 5 (Summer 1998).

Krąkowski, Elie, 'Afghanistan and Beyond: the Strategy of Dismemberment'. *The National Interest*, Vol. 7 (Spring 1987).

Kulesza, Eva, 'Russia's Afghan Syndrome'. *European Journal of International Affairs*, (Fall 1989).

Laber, Jeri and Barnett R Rubin, *A Nation is Dying: Afghanistan Under the Soviets 1979-87* . Evanston, Il: Northwestern University Press, 1988.

Latynski, Maya and S Enders Wimbush, 'The Mujahideen and the Russian Empire' *National Interest,* No. 11 (Spring 1988).

Leitenberg, Milton, 'United States Foreign Policy and the Soviet Invasion of Afghanistan'. *Arms Control,* Vol. 7, No. 3 (December 1986).

Lewis, John W, *Peasant Rebellions and Communist Revolutions in Asia.* Stanford; Stanford University Press, 1974.

Magnus, Ralph H (ed.), *Afghan Alternatives.* New Brunswick, N.J.: Transaction Books, 1985.

_______ ,'Tribal Marxism: The Soviet Encounter with Afghanistan'. *Conflict,* Vol. 21, Nos. 2/3/4 (1983).

Majrooh, S B and S M Y Elmi, *The Sovietization of Afghanistan.* Peshawar: Afghan Information Centre, 1986.

Male, Beverly, *Revolutionary Afghanistan.* London: Croom Helm, 1982.

Maley, William, 'Political Legitimation in Contemporary Afghanistan'. *Asian Survey,* Vol. 27 (June 1987).

Malik, Hafiz, (ed.), *Soviet-American Relations with Pakistan, Iran and Afghanistan.* New York: St. Martin's Press, 1987.

Manzar, A M, *Red Clouds Over Afghanistan.* Islamabad: Institute of Policy Studies 1980.

Matveyev, Vikenti, 'The Soviet Viewpoint'. *Africa Report,* Vol. 21, No. 1 (January-February 1976).

Maududi, Syed Abul Ala, *A Short History of the Revivalist Movement in Islam,* translated from Urdu by al-Ashari, Lahore: Islamic Publications, 1972.

_______ , *First Principles of Islamic State* translated from Urdu and edited by Khurshid Ahmad, Lahore: Islamic Publications, 1974.

Mayer, A C, 'Pir and Murshid: An Aspect of Religious Leadership in West Pakistan'. *Middle Eastern Studies,* Vol. 3, No. 2 (1967).

McCagg, William O, Jr., and Silver, Brian, (eds.) *Soviet Asian Ethnic Frontiers.* New York: Pergamon Press, 1979.

Medvedev, Roy, 'The Afghan Crisis'. *New Left Review,* No. 121 (1980).

Miller, Norman, and Aya, Roderick, (eds.), *National Liberation: Revolution in the Third World.* New York: Free Press, 1971.

Mirror of Jihad (Peshawar), 'Aim and Goals of Jamiat-i-Islami Afghanistan'. *Mirror of Jehad,* Vol. 1, No. 1 (1982).

_______ , 'A brief biography of Professor Burhanuddin Rabbani, the Revolutionary Leader of Jamiat-i-Islami Afghanistan'. *Mirror of Jehad,* Vol. 1, No. 1 (1982).

_______ , Editorial: 'What Is Jehad and Who Is a Mujahid?' *Mirror of Jehad,* Vol. 1, No. 3 (1982).

Misdaq, Nabi, 'Traditional Leadership in Afghan Society and the Issue of National Unity'. *Central Asian Survey,* Vol. 9, No. 4 (1990).

Misra, K P, (ed.), *Afghanistan Crisis.* New Delhi: Vikas Publishing, 1981.

Monks, Alfred L, *The Soviet Intervention in Afghanistan.* Washington and London: American Enterprise Institute for Public Policy Research, 1981.

Moore, Barrington, Jr., *Social Origins of Dictatorship and Democracy: Lord and Peasant in the Making of the Modern World.* Boston: Beacon Press, 1966.

Mukherjee, Sadhan, *What is Happening in Afghanistan.* New Delhi: Communist Party of India, Publication No. 9, July 1981.

Muradov, Gulam, 'National Democratic Revolution in Afghanistan: A Soviet View'. *Journal of South Asian and Middle Eastern Studies,* Vol. 6, No. 1 (Fall 1982).

Murarka, Dev, 'Afghanistan: The Russian Intervention; A Moscow Analysis'. *Strategic Digest* (January 1982).

Naby, Eden 'Ethnic Factors in Soviet-Afghan Relations'. *Asian Survey,* Vol. 20 (1980).

———, 'Islam within the Afghan Resistance'. *Third World Quarterly,* Vol. 10, No. 2 (April 1988).

———, 'The Concept of Jihad in Opposition to Communist Rule: Turkestan and Afghanistan'. *Studies in Comparative Communism,* Vol. 19, No. 3/4 (Autumn/Winter 1986).

National Islamic Front of Afghanistan, *Manifesto.* Peshawar: n.d.

Nayar, Kuldip, *Report on Afghanistan.* New Delhi: Allies Publishers, 1981.

Negaran, Hannah, (Pseud.) 'Afghanistan: A Marxist Regime in a Muslim Society'. *Current History,* Vol. 76, No. 446 (April 1979).

———, 'The Afghan Coup of April 1978: Revolution and International Security'. *Orbis,* Vol. 23, No. 1 (Spring 1979).

———, 'Post-Soviet Afghanistan: the Position of the Minorities'. *Asian Survey,* Vol. 29 (November 1989).

———, 'Revolution and Revolt in Afghanistan'. *The World Today,* Vol. 35, No. 11 (November 1979).

Newell, Richard S, *The Politics of Afghanistan.* Ithaca: Cornell University Press, 1972.

———, 'Post-Soviet Afghanistan: The Position of the Minorities'. *Asian Survey,* Vol. 29 (November 1989).

Newman, Joseph, Jr., 'The Future of Northern Afghanistan'. *Asian Survey,* Vol. 28, (July 1988).

Nomani, Maulana Javed, 'Opponents of Saur Revolution are Enemies of Islam'. *Secular Democracy,* Vol. 15, No. 2 (February 1982).

Norvell, Douglas G, *Agricultural Credit in Afghanistan: A Review of Progress and Problems from 1954 Until 1972.* Kabul: United States Agency for International Development, Agricultural Division, 1972.

On the Saur Revolution , Kabul: Political Department of the People's Armed Forces of Afghanistan, Government Printing Press, 22 May 1978.

Pakistan Progressive, 'Interview with an Afghan Marxist'. *Pakistan Progressive,* Vol. 3, No. 2 (1980).

Panjsheri, Dastagir 'A Glimpse at the History of Revolutionary Afghanistan'. *Democratic Republic of Afghanistan Annual,* (1978-9).

Paul, Jim, 'The Khalq Failed to Comprehend the Contradictions of the Rural Sector: Interview with Feroz Ahmad'. *MERIP Reports*, Vol. 10, No. 6 (July-August 1980).

Payind, Alam, 'Soviet-Afghan Relations from Co-operation to Occupation'. *International Journal of Middle East Studies*, Vol. 21, No. 1 (February 1989).

PDPA's Platform. Kabul Khalq, 11 April 1966.

Peters, Rudolph, *Islam and Colonialism: The Doctrine of Jihad in Modern History*. The Hague and New York; Mouton, 1979.

_______ , *Jihad in Mediaeval and Modern Islam*. Leiden: E J Brill, 1977.

Petkov, Boris, *Afghanistan Today: Impressions of a Journalist*. New Delhi: Sterling Publishers, 1983.

Phillips, James A, 'Moscow's Poison War: Mounting Evidence of Battlefield Atrocities'. *Backgrounder*, (February 5, 1982).

Pickard, Sir Cyril, 'Afghanistan: Difficult Decisions for the West'. *The Round Table*, No. 278 (April 1980).

Pipes, Richard, *U.S.-Soviet Relations in the Era of Detente*, Boulder, Colo.: Westview Press, 1971.

Pope, Ronald R, 'Miscalculation in Soviet Foreign Policy (With Special Reference to Afghanistan)'. *Crossroads*, No. 8 (Winter/Spring 1982).

Poullada, Leon B, 'Afghanistan and the United States: The Crucial Years'. *Middle East Journal*, Vol. 35, No. 2 (Spring 1981).

_______ , *Reform and Rebellion in Afghanistan, 1919-1929: King Amanullah's Failure to Modernize a Tribal Society*. Ithaca: Cornell University Press, 1973.

Prokhanov, Alexander, 'Why the Soviets failed in Afghanistan'. *Globe* (Karachi), (June-July 1989).

Rahman, Fath-ur, and Bashir A Qureshi, *Afghans Meet Soviet Challenge*. Peshawar: Institute of Regional Studies, 1986.

Rais, Rasul Bakhsh, 'Afghanistan after the Soviet Withdrawal'. *Current History*, Vol. 91, No. 563 (March 1992).

_______ , 'Afghanistan and Regional Security After the Cold War'. *Problems of Communism*, Vol. 61 (May-June 1992).

_______ , 'The Afghan Conflict: Is a Political Settlement Possible?'. *Bulletin of Peace Proposals*, Vol. 22, No. 3 (September 1991).

Ramji, Lal, 'Afghanistan Crisis and the Role of the Muslim Countries'. *International Review of History and Political Science*, Vol. 19, No. 2 (May 1982).

Rashid, Abdul, 'Behind the Soviet Pullout: the Specter of Afghanistan's Partition'. *Strategic Review*, Vol. 16, No. 2 (Spring 1988).

Ratnam, Perala, *Afghanistan's Uncertain Future*. New Delhi: Tulsi Publishing, 1981.

Reisman, W Michael, 'The Resistance in Afghanistan is Engaged in a War of National Liberation'. *American Journal of International Law*, Vol. 81, No. 4 (October 1987).

Republic of Afghanistan, *National Reconciliation: Documents.* Kabul: State Printing House, January, 1987.

Rizvi, Gowher, 'The Endgame in Afghanistan'. *The World Today*, Vol. 47, No. 2 (February 1991).

Rizvi, Hasan Askari, 'Geneva Parleys on Afghanistan'. *Pakistan Horizon*, Vol. 39, No. 1 (1986).

Rogers, Tom, 'Afghan Refugees and the Stability of Pakistan'. *Survival*, Vol. 29, No. 5 (September-October 1987).

———, 'Afghan Refugees in Pakistan: a Factor in Political Instability'. *Ethnic Studies Report*, Vol. 8, No. 1 (January 1990).

Rogers, Tom and Anthony Hyman, 'Afghans in Exile'. *Conflict Studies*, Vol. 22 (August 1987).

Roy, Olivier, 'Afghanistan: Back to Tribalism or On to Lebanon?' *Third World Quarterly*, Vol. 11, No. 4 (October 1989).

———, *Islam and Resistance in Afghanistan.* Cambridge: Cambridge University Press, 1986.

Rubin, Barnett R, 'Afghanistan: Political Exiles in Search of a State'. *Journal of Political Science*, Vol. 18 (1990).

———, 'Afghanistan: the Next Round'. *Orbis*, Vol. 33, No. 1 (Winter 1989).

———, 'Lineage of the State in Afghanistan'. *Asian Survey*, (November 1988).

———, 'Political Elite in Afghanistan: Rentier State Building, Rentier State Wrecking'. *International Journal of Middle East Studies*, No. 24 (1992).

———, 'The Fragmentation of Afghanistan'. *Foreign Affairs*, Vol. 68, No. 5 (Winter 1989-90).

———, 'The Old Regime in Afghanistan: Recruitment and Training of a State Elite'. *Central Asian Survey*, Vol. 10, No. 3 (1991).

Rubinstein, Alvin Z, ' Afghanistan: Speculations on a National Tragedy'. *Orbis*, Vol. 30, No. 4 (Winter 1987).

———, *Soviet Policy Toward Turkey, Iran and Afghanistan: The Dynamics of Influence.* New York: Praeger Publishers, 1982.

Rupert, James, 'Afghanistan's Slide Toward Civil War'. *World Policy Journal*, Vol. 6, No. 4 (Fall 1989).

Saikal, Amin, 'Afghanistan: A New Approach'. *Journal of Democracy*, Vol. 3, No. 2 (April 1992).

———, 'Russia and Afghanistan; A Turning Point?'. *Asian Affairs*, Vol. 20, (old series 76), part II (June 1989).

Salzman, Philip, 'Ideology and Change in Tribal Society'. *Man*, No. 13 (1978).

Sardar, Riffat, 'The Afghan Crisis: Seven Years Later'. *Journal of South Asian and Middle Eastern Studies*, Vol. 9, No. 2 (Winter 1985).

Sayyaf, Abdul-Rab Rasul, Professor, *The Two Statements:* Place of publication and publisher n.a., 1980.

Schurmann, H F, *The Mongols of Afghanistan: An Ethnography of the Mongols and Related People in Afghanistan*; The Hague: Mouton, 1962.

Shah, Iqbal Ali, *Afghanistan of the Afghans.* London: Diamond, 1928.

_______ , *Modern Afghanistan.* London: Low, Marston, 1939.

Shahi, Agha, *Pakistan's Security and Foreign Policy.* Lahore: Progressive Publishers, 1988.

_______ , 'The Geneva Accords'. *Pakistan Horizon,* Vol. 41, No. 3 (July 1988).

Shahrani, M Nazif, *The Kirghiz and Wakhi of Afghanistan: Adaptation to Closed Frontiers.* Seattle: University of Washington Press, 1979.

Shahrani, M N and Robert L Canfield, (eds.), *Revolutions and Rebellions in Afghanistan.* Berkeley: Institute of International Studies, University of California, 1984.

Shalinsky, Audrey C, 'Central Asian Emigres in Afghanistan; Problems of Religious and Ethnic Identity'. New York; Asia Society, Afghanistan Council Occasional Paper No. 19, 1979.

_______ , 'The Significance of Islam in Pre -1978 Northern Afghanistan: An Urban Uzbek Example'. *Central Asian Survey,* Vol. 9, No. 4 (1990).

Sheikh, Ali T, 'Afghanistan: an Appraisal of Soviet Withdrawal Strategies'. *Pakistan Horizon,* Vol. 42, Nos. 3/4 (October 1989).

Shorish, Mobin, 'The Impact of the Kemalist "Revolution" on Afghanistan'. *Journal of South Asian and Middle Eastern Studies,* Vol. 7, No. 3 (1984).

Shultz, Richard H, Jr., 'Low-intensity Conflict: Future Challenges and Lessons from the Reagan Years'. *Survival,* (July/August 1989).

Sierakowska-Dyndo, Jolanta, 'The State in Afghanistan's Political and Economic System on the Eve of the April 1978 Coup'. *Central Asian Survey,* Vol. 9, No. 4 (1990).

Silber, Irwin, *Afghanistan—The Battle Line Is Drawn.* San Francisco; Line of March Publications, 1980.

Sinha, P B,'The Afghan Revolution and After'. *Foreign Affairs Reports,* (New Delhi), Vol. 28, No. 7 (July 1979).

Skocpol, Theda, *States and Social Revolutions.* Cambridge: Cambridge University Press, 1979.

Sliwinski, Marek K, 'On the Routes of "Hijrat".' *Central Asian Survey,* Vol. 8, No. 4 (1989).

_______ , 'Afghanistan: The Decimation of a People'. *Orbis,* Vol. 33, No. 1 (Winter 1989).

Smith, Harvey H, et al. *Area Handbook for Afghanistan,* 4th ed. Washington, D.C.: Government Printing Office, 1973.

Social Sciences Today, Afghanistan: Past and Present. Moscow: USSR Academy of Sciences, Oriental Studies in the USSR, 1981.

Solarz, Stephen J, 'When to Intervene'. *Foreign Policy,* No. 63 (Summer 1986).

Spain, David W, *The Pathan Borderland.* The Hague: Mouton, 1963.

Steele, Jonathan, 'Moscow's Kabul Campaign'. *MERIP Reports,* No. 141 (July-August 1986).

Stewart, Rhea Talley, *Fire in Afghanistan, 1914-29: Faith, Hope and the British Empire.* New York: Doubleday, 1973.

Stork, Joe, 'U.S. Involvement in Afghanistan'. *MERIP Reports,* No. 89 (1980).

Strmecki, Marin, 'Gorbachev's New Strategy in Afghanistan'. *Strategic Review*, Vol. 15, No. 3 (Summer 1987).

Svetosarov, Vladimir, 'Afghan Revolution'. *International Affairs* (Moscow). Vol. 15 (May 1983).

_______, 'Afghanistan: Strengthening Revolutionary Gains'. *Asia and Africa Today* (Moscow), No. 2 (1983).

Taper, Nancy, 'Pashtun Nomad Women in Afghanistan'. *Asian Affairs*, Vol. 7, No. 2 (1977).

_______, 'The Advent of Pashtun Maldars in Northwestern Afghanistan'.*Bulletin of the School of Oriental and African Studies*, Vol. 36, No. 1 (1973).

Taper, Richard, *The Conflict of Tribe and State in Iran and Afghanistan*. London: Croom Helm, 1983.

Tarzi, Shah M, 'Politics of the Afghan Resistance Movement: Cleavages, Disunity, and Fragmentation'. *Asian Survey*, Vol. 31 (June 1991).

Tears, Blood and Cries: Human Rights in Afghanistan Since the Invasion 1979-84. New York: Helsinki Watch, 1984.

Thomas, Raju C C, 'Afghanistan Crisis and South Asian Security'. *Journal of Strategic Studies*, Vol. 4, No. 4 (December 1981).

Tilly, Charles, 'Does Modernization Breed Revolution?' *Comparative Politics*, Vol. 5, No. 3 (1973).

Trousdale, William, (ed.), *War in Afghanistan, 1879-80: The Personal Diary of Major General Sir Charles Metcalfe MacGregor*. Detroit: Wayne State University Press, 1985.

US Congress, House Committee on Foreign Affairs, Subcommittee on Europe and the Middle East. *Hearing: East-West Relations, United States-Soviet Relations*. 24 and 30 January 1980, 96th Congress, 2nd Session. Washington, D.C.: US Government Printing Office, 1980.

US Department of State, *Soviet Invasion of Afghanistan*. Special Report No. 70, April 1980.

_______, *The Soviet Dilemma in Afghanistan*. Washington, D.C.: June 1981.

Uberoi, J P Singh, 'District Administration in the Northern Highlands of Afghanistan'. *Sociological Bulletin*, Vol. 17, (March 1968).

Ulam, Adam B, *Dangerous Relations: The Soviet Union in World Politics*. New York: Oxford University Press, 1983.

_______, *Expansion and Coexistence: The Soviet Foreign Policy, 1917-73*. 2nd ed. New York: Holt, Rinehart and Winston, 1974.

_______, *The Rivals: America and Russia Since World War II*. New York: Viking Press, 1971.

Urban, Mark, *War in Afghanistan*. New York: St. Martin's Press, 1988.

Valenta, Jiri, 'From Prague to Kabul: The Soviet Style of Invasion'. *International Security*, Vol. 5, No. 2 (Fall 1980).

_______, 'The Soviet Invasion of Afghanistan: The Difficulty in Knowing Where to Stop'. *Orbis*, Vol. 24, No. 2 (1980):.

Vedik, Vet Pratab, 'Interview with Babrak Karmal'. *Strategic Digest* (July 1981).

Verduin, Leonard, *The Reformers and Their Step-Children.* Grand Rapids, Mich.: Baker, 1964.

Volkov, Y, et al., *The Truth about Afghanistan: Documents, Facts and Eyewitness Reports.* Moscow: Novosti Press, 1980.

Wakil, Abdul, 'Afghanistan Solution, the First Track'. *Orient,* Vol. 30, No. 3 (September 1989).

_______ , 'Iran's Relations with Afghanistan after the Islamic Revolution'. *Orient,* Vol. 32, No. 1 (March 1991).

Wardak, Gen. Rahim, 'A Synopsis of Marxist Tragedy in Afghanistan'. *The Front: The Mujahideen Monthly.* Peshawar, (March 1978).

Watt, W Montgomery, *Islamic Political Thought: The Basic Concepts.* Edinburgh: Edinburgh University Press, 1968.

Weinbaum, Marvin G, 'Pakistan and Afghanistan: the Strategic Relationship'. *Asian Survey,* Vol. 31 (June 1991).

_______ , 'The Politics of Afghan Resettlement and Rehabilitation'. *Asian Survey,* Vol. 29, (March 1989).

_______ , 'Legal Elites in Afghan Society'. *International Journal of Middle Eastern Studies,* Vol. 12 (1980).

Wilber, D, *Afghanistan: Its People, Its Society, Its Culture.* New Haven: Human Relations Area Files Press, 1962.

Wilbur, D N, 'The Structure and Position of Islam in Afghanistan'. *Afghanistan,* Vol. 10 (1955).

Wirsing, Robert G, 'Pakistan and the War in Afghanistan'. *Asian Affairs, An American Review.* Vol. 14, No. 2 (Summer 1987).

Yadav, Sanjay Singh, 'Failed Great Power War and the Soviet Retreat from Afghanistan'. *Comparative Strategy,* Vol. 8, No. 4 (1989).

Yapp, Malcolm E, 'Disturbances in Afghanistan, 1839-42'. *Bulletin of the School of Oriental and African Studies,* Vol. 25 (1962).

_______ , 'Disturbances in Western Afghanistan, 1839-41'. *Bulletin of the School of Oriental and African Studies,* Vol. 26 (1963).

_______ , 'The Revolutions of 1841-2 in Afghanistan'. *Bulletin of the School of Oriental and African Studies,* Vol. 27 (1964).

Yetiv, S A, 'How the Soviet Military Intervention in Afghanistan Improved the US Strategic Position in the Persian Gulf'. *Asian Affairs, An American Review,* Vol. 17, No. 2 (Summer 1990).

Yusufzai, Rahimullah, 'PDPA: Khalq-Parcham Struggle for Power', *Regional Studies,* Vol. 2, No. 2 (Spring 1984).

Zeary, Saleh Mohammad and Nur Ahmad Nur, 'Afghanistan; the revolution marches on'. *Soviet Review,* Vol. 20 (12 May 1983).

_______ , 'Afghanistan: The beginning of a new era'. *World Marxist Review,* Vol. 22, No. 1 (January 1979).

Zekrya, Mir-Ahmad B, 'Planning and Development in Afghanistan: A Case of Maximum Foreign Aid and Minimum Growth'. Ph.D. dissertation, Johns Hopkins University, 1976.

Ziring, Lawrence, 'Soviet Policy on the Rim of Asia: Scenarios and Projections'. *Asian Affairs*, Vol. 9, No. 3 (Jan-Feb. 1982).

Zonis, Marvin, 'Middle East Responses'. *Orbis*, Vol. 30, No. 4 (Winter 1987).

Index